William Healey Dall

Nomenclature in zoology and botany

A report to the American Association for the Advancement of Science at the

Nashville meeting, August 31, 1877

William Healey Dall

Nomenclature in zoology and botany
A report to the American Association for the Advancement of Science at the Nashville meeting, August 31, 1877

ISBN/EAN: 9783741198670

Manufactured in Europe, USA, Canada, Australia, Japa

Cover: Foto ©Klaus-Uwe Gerhardt /pixelio.de

Manufactured and distributed by brebook publishing software (www.brebook.com)

William Healey Dall

Nomenclature in zoology and botany

NOMENCLATURE

IN

ZOOLOGY AND BOTANY.

A REPORT TO THE AMERICAN ASSOCIATION FOR THE ADVANCEMENT OF
SCIENCE AT THE NASHVILLE MEETING, AUGUST 31, 1877.

BY

W. H. DALL,

UNITED STATES COAST SURVEY.

SALEM:
PRINTED AT THE SALEM PRESS.
DECEMBER, 1877.

APPENDIX.

CONTENTS.

REPORT OF THE COMMITTEE ON ZOOLOGICAL NOMENCLATURE TO
SECTION B, OF THE AMERICAN ASSOCIATION FOR THE ADVANCE-
MENT OF SCIENCE, AT THE NASHVILLE MEETING, AUGUST 31,
1877.[1]

THE undersigned was appointed at the last meeting a committee
of one, to obtain an expression of opinion from the working naturalists of America, in regard to the nature of a set of rules for
facilitating the decision of questions relating to nomenclature,
and which might be adopted or recommended by the Section.

In accordance with the understanding and resolutions of the
Section, by which this duty devolved upon him, your Reporter
prepared a circular which was printed under the supervision of the
Permanent Secretary and circulated by the Smithsonian Institution, a copy of which is appended to this report.

The recently published "Naturalists' Directory," afforded the
addresses of nearly all the publishing naturalists, in the discrimination of whom, valuable aid was received from Messrs. Scudder
and LeConte.

The circular was sent to all who, within the last five years,
might be included under the designation of publishing naturalists,
and of whom the address could be obtained. This list included
about eighty-five names, from a number of which (for various
reasons), a response was hardly anticipated. They were used,
however, in order that the fullest opportunity might be afforded to
all those who might desire to express an opinion.

The responses received to date (Aug. 14, 1877), are forty-five
in number. While a few honored names, to whose views all would
attribute due weight, are not on the list, yet it includes most of
those whose contributions are familiar in the Proceedings of Amer-

[1] By vote of the Section this report was referred to the Standing Committee, who
directed that the report be printed and further action upon it deferred until the next
meeting of the Association.

ican Scientific Societies, and an unquestionable majority of the best working-naturalists of the country. The views of several of those from whom no response was received, have been incorporated in the appendix by means of citations from their works.

The queries contained in the circular relate chiefly to certain points, concerning which a diversity of opinion has existed among naturalists ; the general principles of nomenclature not being in question. The responses are divided into affirmative, negative and doubtful, while in individual cases some queries received no response. The answers classed as doubtful, comprise those which by their tenor indicated that the purport of the particular query had not been clearly understood, and some in which the person replying avowed his inability to express a preference for any one of several modes of proceeding.

The gratifying unanimity which is exhibited in the responses to certain of the more important and clearly defined questions at is-sue, indicates that a thorough study of the more complicated questions by the light of the general principles of nomenclature, would result in a practical agreement on the part of American Naturalists in relation to nearly all the matters in debate.

It is evident from the responses of naturalists, that their opinion is generally adverse to any attempt to limit, by arbitrary rules, the right of publication in the most convenient direction, and against any statute of limitations in regard to scientific names. This seems to be in accord with the principles of justice, equity and general usage in nomenclature, though at times, inconvenient in its results. It may be confidently expected that the majority of authors, when their attention has been drawn to it will, for their own interests as well as that of science, avoid in future publica-tions, the methods (or want of method), which in the remote past sowed so many germs of present difficulty.

It was thought by your Reporter that the best interests of science would be served, and the whole question be more clearly presented for future decision, by preparing such an appendix as that which forms the chief bulk of this report as now submitted.

This appendix consists of a resumé of all the principles and rules of nomenclature as heretofore set forth by the chief authori-ties on that subject, with the diverse views of different authors concerning each proposition appended to it and authenticated by their initials according to the accompanying list. In most cases, a statement of the bearing of the responses to the circular upon each rule is included in the discussion following the rule. There are also some comments for which the writer is responsible.

A serious mistake appears to have been committed at the outset by divorcing Zoological from Botanical nomenclature, as was done by the committee of the British Association. The signal success which has attended the efforts of botanists to unify their nomenclature, when compared with the confusion reigning in some departments of zoology, is sufficient proof of this. Your Reporter, therefore, has endeavored to combine both in a symmetrical manner in one general whole. The principles and almost all the details are essentially identical in both. In performing this labor great dependence has been placed on the admirable memoir of De Candolle, and the writer is indebted to Dr. Asa Gray for kind suggestions in regard to doubtful points.

The rules and examples for the proper construction of compounds and derivatives from classical roots, for properly Latinizing vernacular terms, and the table of equivalent letters in Greek and Latin, have been amplified beyond the extent previously attempted by any writer on nomenclature, and also submitted for criticism to a competent professor of languages to insure accuracy. It is believed that this feature will prove useful to those naturalists not especially familiar with the classical languages.

While it has seemed necessary to criticise, in one or two matters of detail, the rules put forward by the two committees of the British Association, yet the wisdom and advisability of their recommendations in general must be fully conceded.

While the forms of expression, used by authors cited, have been frequently somewhat modified in translation or for harmony of diction, great care has been taken not in any way to modify the opinions expressed by each writer. For the Reporter's comments he is of course solely responsible. The few suggestions which he has ventured to make are open to free criticism; from which and from a study of the broad principles underlying the whole subject, a more satisfactory understanding on the matters in controversy can hardly fail to come about.

It will be readily understood that the labor involved in the preparation of this paper has not been slight, yet it seemed that in no other way could the matters concerned be brought before those interested in a satisfactory and impartial manner.

This report does not form a proposition to be acted on by the
Section, or by the Association in a final manner at the present
time. It is merely an attempt at a complete presentation of the
subject, without which no well advised action by naturalists as a
body can be expected or is to be desired. So far as is known to
the Reporter, no such compilation has hitherto been attempted in
modern times.

Many of the difficulties with which authors are confronted, are
the result of efforts by specialists to remedy confusion in their
particular departments without reflection on the general bearing
of the usages which they have attempted to introduce.

When free discussion, assisted by the light which this report
may tend to throw upon nomenclature, shall have tested the opinions expressed and the reforms which have here and elsewhere
been suggested, a fit time will have arrived to consider a more
concise and definitive code of rules and the subject of their recommendation by the Section. For the present this Report with its
appendices is respectfully submitted.

WM. H. DALL,
Committee.

CIRCULAR.

SMITHSONIAN INSTITUTION, WASHINGTON, D. C.,
February 1, 1877.

DEAR SIR:

AT the Buffalo meeting of the American Association for the Advancement of Science, 1876, a report from the Committee on Zoological Nomenclature was presented by the Chairman, Dr. LeConte, and after discussion it was decided that the undersigned be appointed a committee of one for the purpose of obtaining an expression of opinion from the working naturalists of the United States and Canada, in regard to the nature of a set of rules for facilitating the decision of questions relating to nomenclature, and which might be adopted and recommended by the Association. For this purpose it was decided to issue circulars. It was further understood that these circulars should be prepared, distributed, and, upon their return, a report upon the result prepared by the undersigned, to be presented at the Nashville meeting in 1877; these proceedings to be preliminary to consideration of, and definite action on the subject by the Association, if action be decided to be desirable.

In order to place some reasonable limit on the number to be distributed, it was decided that the circular should be sent only to those naturalists who have, during the last five years published descriptive zoological or botanical papers.

The question under consideration being one in which every naturalist must be more or less concerned, it is confidently expected that the persons who may receive the circular, will devote to it the small amount of time required to fulfil its purpose, in order that a full expression of opinion may be obtained. In order that as little writing as possible shall be required, the questions have been so framed as in most cases to call for only a categorical answer. This method is also absolutely essential to prevent ambiguity.

You are, therefore, requested to insert in the spaces left for the purpose in the accompanying circular, such answers to the questions of which it is composed, as are in your opinion most desirable and expedient. Having done this, you are further requested to sign and mail the circular to the undersigned as promptly as possible, in order that the preparation of the report may not be delayed.

I remain very respectfully,

WM. H. DALL,

Committee.

(13)

NOTE.

The questions with which the working Naturalist is most frequently brought face to face — and in the decision of which so much trouble is experienced and such diverse opinions are elicited — are chiefly those which involve the right of any one of several names to be considered as properly proposed and entitled to take precedence of others, provided its priority in time of application be established.

The rule that names (otherwise unexceptionable) which are prior in date, are to be accepted in nomenclature to the exclusion of all others, is conceded by all naturalists.

The rules recommended by the Committee on Nomenclature of the British Association for the Advancement of Science, have been generally adopted: though in certain details they are regarded by many naturalists as defective. Nevertheless they have largely contributed to that uniformity which is so desirable in the matter of nomenclature.

It has been thought that a similar recommendation on the part of the American Association might reach many who are not conversant with the British rules and tend to produce in the works of the rising generation of American Naturalists a similarly beneficial agreement.

The differences of opinion which have arisen, are chiefly in matters of detail and intrinsically of very slight importance.

One of the most serious in its effect upon nomenclature is that in regard to what names shall be considered as really binomial; another as to what is necessary to definitely establish a name in order that if prior to any other it may be accepted as properly proposed; and most of all as to the date to be adopted as that of the beginning of binomial nomenclature. This latter question, as to facts, on the authority of De Candolle, stands as follows :

A series of rules for nomenclature was to some extent foreshadowed by Linnæus in his *Fundamenta Entomologia* of 1736. These rules were first definitively proposed in the *Philosophia botanica*, which appeared in 1751. These rules, however, related almost exclusively to the generic name or *nomen genericum*. In 1745, he had employed for the first time a specific name (*nomen triviale*) composed of one word, in contradistinction to the polynomial designation of a species (*nomen specificum*) which was previously the rule among naturalists. That which now seems the most happy and important of the Linnæan ideas, the restriction of the specific name as now understood, seems to have been for a long time only an accessory matter to him, as the *nomina trivialia* are barely mentioned in his rules up to 1765.

In 1753, in the *Incrementa botanices*, while expatiating on the reforms which he had introduced into the science, he does not even mention the binominal nomenclature. In the *Systema Naturæ*, Ed. X, 1758, for the first time the binominal system is consistently applied to all classes of animals and plants (though it had been partially adopted by him as early as 1745), and hence many naturalists have regarded the tenth edition as forming the most natural starting point. The system being of slow and intermittent growth, even with its originator, an arbitrary starting point is necessary. In the twelfth edition (1766-68), numerous changes and re- forms were instituted, and a number of his earlier specific names were arbitrarily changed. In fact, Linnæus never seems to have regarded spe- cific names as subject to his rules.

The last was recommended by the British Committee as the starting point. They have since, however, receded to the extent of admitting to recognition some ichthyological works printed between the dates of the tenth and twelfth editions.

QUESTIONS TO WHICH AN ANSWER IS DESIRED.

I. What date shall be taken as the commencement of the binomial era in nomenclature? For Ed. X, 18. Ed. XII, 17. 1736, 1. Botanists, 1753, 2. No answer, 7.

II. Shall phrases composed of two words which may appear in the publications of naturalists whose works preceded, or who did not in such works adopt the binomial system of nomenclature, be considered as binomial names? No, 32. Yes, 5. No answer or doubtful, 8.

III. If so, shall the first word of the said phrase be entitled to recognition as a generic name? No, 32. Yes, 5. No answer or doubtful, 8.

IV. If an author has not indicated his adoption of the binomial system by discarding all polynomial names in a given work, are any of his names therein entitled to recognition otherwise than in bibliography? No, 18. Yes, 18. Doubtful, 4. No answer, 5.

> *Example.* Da Costa in his work on the Conchology of Great Britain, varies from binomial to polynomial in his designations of species, and some of his " generic " names contain two or three words, while others apparently conform to the Linnæan system. Should any of these names be retained ?

V. Does the reading of a paper before a scientific body constitute a publication of the descriptions or names of animals or plants contained therein ? No, 39. Doubtful, 2. Yes, 4.

VI. Is a name in the vernacular of the publishing author, or a vernacular rendering from a classical root unaccompanied by a Latin or Greek form of the name, entitled to recognition except in bibliography? No, 36. Doubtful, 2. Yes, 4. No answer, 3.

VII. Is a name applied to a group of species without a specification of any character possessed by them in common (that is, without any so-called generic diagnosis or description), entitled to recognition as an established generic name by subsequent authors? No, 38. Doubtful, 3. Yes, 3. No answer, 1.

VIII. Is a generic name applied to a single (then or previously) described species without a generic diagnosis or description of any kind, entitled to recognition as above, by subsequent authors? No, 37. Doubtful, 3. Yes, 4. No answer, 1.

IX. Is a name, when used in a generic sense, and otherwise properly constituted, subject to have its orthography changed by a subsequent author, on the ground that a proper construction from its classical roots would result in a different spelling? No, 21, Doubtful, 3. Yes, 19. No answer, 2.

X. If the previous question be answered in the affirmative, it may be further enquired whether an author has a right to assume that a given name is derived from classical roots, when the author of the name did not so state, and on this assumption to proceed to change the said name to make it agree with the assumed proper construction in any case? and especially when by the asserted reformation the generic name becomes identical with one previously proposed for some other animal or plant, and hence will fall into synonymy? No, 25. Doubtful, 2. Yes, 6. No answer, 12.

Example. Schumacher described a genus which he called *Paxydon*, giving no derivation. A subsequent author described a genus *Pachydon*, giving the derivation. A third writer assumed that Schumacher's name had the same derivation as *Pachydon*, and that both, if correctly written, would be *Pachyodon*. The last mentioned then proposed a new name for *Pachydon*, which he had thus made to appear preoccupied. Was this allowable? No, 26. Doubtful, 3. Yes, 8. No answer, 8.

XI. Should a generic name, otherwise properly constituted, but derived from the specific name of its typical species, or similar to that of one of the species included under it, be rejected on that account? No, 40. Doubtful, 4. Yes, 1.

Note. It is proper to state that this is an important question, since Linnæus himself, and others, formed many generic names in this manner, and a large number of such names are currently accepted, especially in botany and among vertebrate animals.

XII. Shall a subsequent author be permitted in revising a composite genus (of which no type was specified when it was described) to name as its type a species not included by the original author of the genus in that latter author's list of species given when the genus was originally described? No, 37. Doubtful, 2. Yes, 5. No answer, 1.

Example. Linnæus described a genus *Chiton* with a very few species. After many species had been described by others, a later author divided the genus into a number of genera, and reserved the name of Chiton (restricted) for a species described many years after the death of Linnæus and belonging to a section of the *Chitonidæ* unknown to Linnæus; while to the Linnæan chitons he gave new appellations.

XIII. When an old genus without a specified type has been subdivided by a subsequent author, and one of the old species retained and specified to be the type of the restricted genus bearing the old name, — is it competent for a third author to discard this and select another of the original species as a type, when by so doing changes are necessitated in nomenclature? No, 39. Doubtful, 4. No answer, 2.

XIV. Shall an author be held to have any greater control over or greater privileges with relation to names of his own proposing, after the same have been duly published, than any other subsequent author? No, 40. Doubtful, 2. Yes, 2. No answer, 1.

XV. For instance, when an author describes a genus and indicates a species as its type, is it allowable for him subsequently to substitute any other species as a foundation for his genus, or to use the original type as a foundation for another new genus? No, 38. Doubtful, 1. Yes, 2. No answer 4.

XVI. If an author describes a genus and does not refer to it any then or previously described existing species, can the genus be taken as established? No, 33. Doubtful, 7. Yes, 1. No answer, 4.

XVII. If an author applies a specific name to an object without referring it to some then or previously described genus, is the specific name entitled to recognition by subsequent authors? No, 33. Doubtful, 4. Yes, 7. No answer, 1.

XVIII. When a generic name has lapsed from sufficient cause into synonymy, should it be thenceforth entirely rejected from nomenclature? or should it still be applicable to any new and valid genus? Reject, 19. Accept, 23. Doubtful, 1. No answer, 2.

XIX. Should a name which has been once used in one subkingdom, and has lapsed into synonymy, be considered available for use in any other if not entirely rejected from nomenclature? N o 20. Doubtful, 1. Yes, 18. No answer, 6.

XX. Should a name be liable to be changed or a later one substituted for it, if the original be supposed to be inapplicable or contradictory of the characters of the species or genus to which it was applied? No, 28. Doubtful, 3. Yes, 13. No answer, 1.

Example. A fish without teeth was named *Polyodon* which name had come into use; when a later author substituted *Spatularia* on the ground that *Polyodon* was inapplicable.

XXI. Is it advisable to fix a limit of time, beyond which a name which
has been received without objection during that time shall be
held to have become valid, and no longer liable to change
from the resuscitation of obsolete or uncurrent but actually
prior names? No, 28. Doubtful, 1. Yes, 13. No answer, 3.

XXII. If so, what shall this period be? No answer, 35. The others
range from 10 to 100 years.

XXIII. Should it be permitted to alter, or replace by other and different
appellations, class, ordinal and family names, which owing to
the advance of Science and consequent fluctuation of their
supposed limits have become uncharacteristic? Yes, 30. Or
should these also be rigidly subject to such rules of priority
as might be determined on for generic or specific names? No
answer, 4. Yes, 11.

XXIV. Should or should not absolute certainty of identification be
required before it be permissible to reject a modern and
generally adopted name in favor of a prior but uncurrent
designation? Yes, 38. Doubtful, 2. No answer, 5.

Note. Many of the old descriptions of species sufficient for
identification when few species were known, are entirely in-
sufficient at the present day to distinguish between allied
species. Should, therefore, a modern specific name with a re-
cognizable description be made to yield to an older name
unless the identification can be made beyond any cavil?

XXV. Is it desirable to adopt any classification of periodical literature
by which certain exclusive channels for publication of descriptive papers in Natural History shall be designated for use by
authors who desire to secure the rights of priority for new
names proposed by them? No, 26. Desirable but impracticable, 9. Yes, 8. No answer, 2.

Note. An affirmative answer will imply that names which
may be proposed through other than the designated channels,
after the latter shall have been decided upon, shall not be
entitled to recognition in questions of priority.

XXVI. Is it desirable to adopt any analogous rule in relation to the
character or extent of distribution of any independent publication or pamphlet to which it must conform, on pain of
losing its right to recognition? No, 21. Desirable but impracticable, 10. Yes, 14.

Note. If the answer to either or both of the two preceding
questions be affirmative, a note specifying the nature of the
proposed classification or restrictions may be appended to this
list.

XXVII. Should a series of rules be recommended for adoption by the
Association, would you be guided by these recommendations
in cases where they might not agree with your own prefer-
ences? Yes, 29. Yes, with reservations, 15. No, 1.

LIST OF NATURALISTS FROM WHOM REPLIES TO THE CIRCULAR HAVE BEEN RECEIVED.

J. A. Allen, Museum of Comparative Zoology.

W. G. Binney, Burlington, N. J.

Richard Bliss, Jr., Cambridge, Mass.

Dr. Thomas M. Brewer, Boston Society of Natural History.

Dr. P. P. Carpenter, McGill University.

S. F. Clark, Johns Hopkins University.

T. A. Conrad, Philadelphia Academy of Natural Sciences.

Dr. J. G. Cooper, California.

Prof. E. D. Cope, Philadelphia.

W. H. Dall, Smithsonian Institution.

Prof. J. D. Dana, Yale College.

Dr. J. W. Dawson, McGill University.

W. H. Edwards, West Virginia.

S. W. Garman, Museum of Comparative Zoology.

Dr. T. N. Gill, Smithsonian Institution.

Dr. Asa Gray, Harvard University.

A. R. Grote, Buffalo Academy of Sciences.

Dr. Herman Hagen, Museum Comparative Zoology.

Dr. Geo. H. Horn, Philadelphia.

Prof. Alpheus Hyatt, Boston Society of Natural History.

Ernest Ingersoll, New York.

W. P. James, Cincinnati, Ohio.

Prof. D. S. Jordan, Indiana.

Dr. J. L. LeConte, Philadelphia Academy of Natural Sciences.

Dr. Joseph Leidy, Philadelphia Academy of Natural Sciences.

Dr. James Lewis, Mohawk, N. Y.

Theodore Lyman, Museum of Comparative Zoology.

T. L. Mead, New York.

S. A. Miller, Cincinnati, Ohio.

Dr. A. S. Packard, Peabody Academy of Sciences.

F. W. Putnam, Museum of Comparative Zoology.

Prof. C. V. Riley, United States Entomological Commission.

Prof. C. Rominger, State Geologist, Michigan.

Dr. J. T. Rothrock, University of Pennsylvania,

S. H. Scudder, Cambridge, Mass.

Prof. N. S. Shaler, State Geologist of Kentucky.

Herman Strecker, Reading, Pa.

Prof. Cyrus Thomas, United States Entomological Commission.

Geo. W. Tryon, Jr., Philadelphia Academy of Natural Sciences.

P. R. Uhler, Peabody Institute, Baltimore.

Sereno Watson, Harvard University.

Dr. C. A. White, United States Survey of the Rocky Mountain Region.

J. F. Whiteaves, Paleontologist to the Canadian Geological Survey.

Prof. R. P. Whitfield, American Museum of Natural History, N. Y.

Dr. H. C. Yarrow, United States Army.

Two accidentally unsigned.

WORKS REFERRED TO.

Beside the Linnæan canons the following works, beside some not quoted
in the body of this report, have been consulted and are referred to by the
abbreviations which here precede their titles.

A. Nomenclator Zoologicus, auctore L. Agassiz. Soloduri, 1842-46. 4to. (Præ-
    fatio.)

A. Ag. Revision of the Echini; Alexander Agassiz. Cambridge, 1872. 4to. (Prelim-
    inary Remarks on Nomenclature.)

B. A. Rules for Zoological Nomenclature, authorized by Section D of British Asso-
    ciation at Manchester, 1842. 8vo, pp. 26.

B. A. (Modified do.) Report British Assoc., Birmingham, 1865. 8vo, pp. 28. (See
    Verrill.)

Bourg. Methodus conchyliologicus denominationis, par M. J. R. Bourguignat. Paris,
    1860. 8vo, pp. 88.

Br. Index palæontologicus, von Heinrich G. Bronn, vol. III. Stuttgart, 1848.
    (Ueber Nomenclatur, pp. lviii-lxviii.)

DC. Lois de la Nomenclature botanique, rédigées et commentées par M. Alphonse
    de Candolle. Paris, 1867. 8vo, pp. 60.

Gray. Review of Bentham and Hooker, Genera plantarum, by Asa Gray, LL.D., (in)
    Am. Jour. Sci. and Arts, Jan., 1863, p. 134.

G. R. G. From the prefaces to the "List of the Genera of Birds," and to the "Genera of
    Birds," with some additional remarks. London, n. d. 8vo, pp. 10. (Pri-
    vately printed by Geo. R. Gray, about 1870.)

Herrm. Indicis generum malacozoorum primordia, conscripsit A. N. Herrmannsen.
    Vol. I. Cassel, 1846. 8vo. (Leges nomenclationis, pp. vii-xiv.)

LeC.   On some changes in the nomenclature of North American Coleoptera, which have recently been proposed. By John L. LeConte, M.D., (in) Canadian Entomologist, Oct., 1874. 8vo, pp. 185-197.

LeC.   On Entomological Nomenclature, by the same in the same; Nov., 1874. Part I, pp. 201-206. Dec., 1874. Part II, p. 207-210.

Lew.   A discussion of the law of priority in entomological nomenclature, etc., by W. Arnold Lewis. London, 1872. 8vo, pp. 86.

R.   Rules to be submitted to the Entomological Club of the A. A. A. S. (J. L. LeConte, W. Saunders, C. V. Riley), 1876. 8vo, pp. 6.

Scud. Canons of systematic nomenclature applied to the higher groups. S. H. Scudder, (in) Am. Journ. Sci. and Arts (3), III, p. 348. 1872. 8vo.

Scud. Historical sketch of the generic names proposed for butterflies, a contribution to systematic nomenclature, by Samuel H. Scudder. Salem, 1875. 8vo, pp. 293.

Th.   On European Spiders, by T. Thorell; Part I, Observations on Nomenclature. Upsala, 1869. 4to, pp. 88.

V.   Notes on the modified Rules for Zoological Nomenclature, B. A., 1865, by A. E. Verrill, in the Am. Journ. Sci. and Arts, xlviii. July, 1869.

DISCUSSION OF THE SUBJECT OF NOMENCLATURE.

GENERAL PRINCIPLES.

§ I. Natural History cannot progress, nor can the study of its various branches be carried on and properly correlated, without a regular system in nomenclature which shall be recognized and employed by the majority of naturalists of all countries. (DC.)

§ II. The rules for nomenclature must be impartial and founded on motives sufficiently clear and weighty to promote their general comprehension and acceptance. (DC., B. A.)

§ III. The essential principles in everything which relates to nomenclature are (1) the attainment of *fixity* in the designations for organized beings, (2) the avoidance of names or methods of applying names calculated to result in errors or to throw science into confusion, and lastly (3) to avoid the unnecessary creation of names. (DC.)

Other considerations, such as grammatical accuracy in the formation of names, their regularity or euphony, etc., are relatively less important. (DC.)

§ IV. No usage conflicting with the rules and liable to introduce error or confusion can be maintained. When no grave objections of this nature are liable to be raised, it may happen that an ancient usage may be conserved without opposition, but all should carefully guard against the imitation or extension of such practices. In the absence of a rule, or if the application of the rules be doubtful, an established usage may be taken as a proper guide. (DC.)

It is impracticable not to recognize a certain right in *usage*, for, by the maintenance of familiar names and useful forms, clearness and precision are often gained. But it is never desirable to perpetuate a serious error for the small advantage of following a habit. (DC.)

. § V. The principles and forms of nomenclature should be as similar as possible in Botany and Zoology. (DC.)

The manner in which Botany and the different branches of zoology have reached their present state, being far from uniform, and the nature of the organisms treated of being dissimilar, an absolute identity in the application of nomenclature is impracticable even if it were wholly desirable. The fundamental principles, however, and the end to be attained, are the same in both branches of study.

§ VI. Scientific names are of the Latin form or language. When taken from another language they are to be rendered in the manner of Latin words and take a similar termination. If translated into the vernacular for use in a popular or familiar manner the resemblance to the Latin original should be preserved as nearly as possible. (DC., B. A., Th., etc.)

§ VII. Nomenclature comprehends two kinds of names; (1) terms which express the relative value of groups which are comprised each in another. (2) Particular names for each group of organisms known to exist, appellative words which are names and not definitions.

On the Subordination and Designation of Groups belonging to the First Category of § VII.

§ VIII. The following table indicates the two series.

REGNUM VEGETABILE.		REGNUM ANIMALE.	
....	....	Subregnum.	Subkingdom.
Classis.	Class.	Classis.	Class.
Subclassis.	Subclass.	Subclassis.	Subclass.
....	....	Superordo.	Superorder.
Cohors.	Cohort.	Ordo.	Order.
Subcohors.	Subcohort.	Subordo.	Suborder.
....	....	Superfamilia.	Superfamily.
Ordo, familia.	Order or Family.	Familia.	Family.
Subordo, subfamilia.	Suborder.	Subfamilia.	Subfamily.
Tribus.	Tribe.	Tribus.	Tribe.
Subtribus.	Subtribe.	....	....
Genus.	Genus.	Genus.	Genus.
Subgenus.	Subgenus.	Subgenus.	Subgenus.
Sectio.	Section.	Sectio.	Section.
Subsectio.	Subsection.	Subsectio.	Subsection.
Species.	Species.	Species.	Species.
Subspecies, Proles.	Subspecies, Race.	Subspecies.	Subspecies.
Varietas.	Variety.	Varietas.	Variety.
Subvarietas.	Subvariety.	....	....
Variatio.	Variation.	Variatio.	Variation.
Subvariatio.	Subvariation.	....	....
Planta.	Plant.	Animal.	Individual.

The above terms are more or less generally accepted; the relative values being more fully and generally recognized in botany than in zoology. In the literature of the latter branch some of the terms above mentioned are rarely found, though by no means unnecessary for careful discrimination. The term *tribe* in zoology has been used with several different values. In this, as in other respects, the inchoate condition of zoological nomenclature as compared with that of botany is clearly apparent.

§ IX. All individuals belong to a species, all species to a genus, all genera to a family, all families to an order or to a cohort, all orders or cohorts to a class. (DC., B. A.) In some species varieties or variations are recognized, and even more numerous modifications in some cultivated plants; botanical orders are often divided into tribes, and genera into sections. Intermediate groups are distinguished by the prefixes *sub* and *super*, the former being intermediate between the rank of the term to which it is prefixed, and the next lower, the latter between the same term and the next higher.

The botanical terms are not precisely interchangeable with those used in zoology, owing to the differences in the subjects classified and for other reasons; this is especially the case with the terms order and family, in the two series. A special series of terms to indicate *sports, hybrids*, etc., is in use in botany which is exhaustively treated of by DeCandolle.

§ X. The definition of each of these terms or names of groups varies, up to a certain point according to the state of science or the views of the individual writer using them, but their relative rank, sanctioned by usage, cannot be inverted. No classification containing inversions, such as a division of a genus into families or of a species into genera can be admitted. (DC.)

§ XI. In botany subdivisions of a species are usually indicated by figures (1, 2, 3, etc.) and subdivisions of varieties by Greek letters (α, β, λ, etc.) but in zoology it is a common practice to denominate the subspecies by distinct names which are appended to the specific name, as *Ovis ammon* subsp. *crassicornis*.

On the Manner of Designating Particular Groups or those belonging to the Second Category of § VII.

§ XII. Each natural group of animals or plants can have but one valid designation. (DC., B. A., Bourg., etc.) This designation by modern usage is allowed to exist simultaneously in zoology and botany without rendering either application of it invalid, but in either department it can be applied in a valid manner but once.

§ VI should be kept in mind. " We refer solely to the Latin or systematic language of zoology (and botany). We have nothing to do with vernacular designations." (B. A., Th.)

The special application of this principle to genera and species will be treated under their specific heads.

§ XIII. This designation should only be changed for the most important reasons, founded on a thorough knowledge of the facts; or on the necessity of abandoning a denomination on account of its being in conflict with the following rules or with the essential principles of systematic nomenclature. This necessarily follows from § III.

§ XIV. The form, number and arrangement of the names are dependent on the nature of each group, according to the rules which follow.

DENOMINATION OF THE HIGHER GROUPS.

GROUPS OF HIGHER VALUE THAN FAMILIES OR BOTANICAL ORDERS.

§ XV. The names of these groups are taken from some one of the principal characters. They are expressed by single words of Greek or Latin origin in which a certain harmony of form and termination is preserved for groups of similar nature. Ex. *Phanerogamæ, Cryptogamæ; Cephalopoda, Gasteropoda; Raptores, Scansores; etc.* (DC., etc.)

Compounds of Greek and Latin words are inadmissible (Bourg.) Nomina sesquipedalia fugienda sunt (Lin.) In cryptogamic botany ancient names of families such as *Musci, Filices*, etc., have been employed as names of classes or sub-classes. Botanical cohorts or subcohorts are designated by the name of one of their principal families with the termination *ales*. (DC.)

BOTANICAL FAMILIES OR ORDERS.

§ **XVI.** The families (*ordines*) in botany are designated by the name of one of their principal genera, with the termination *aceæ*. Ex. *Rosa, Rosaceæ; Ranunculus, Ranunculaceæ.*

Usage has justified the following exceptions.

1. When the genus from which the name of the family is taken ends in Latin with *ix* or *is* (genitive *icis, idis* or *iscis*) the termination *iceæ*. *ideæ* or *ineæ*, is permitted; as in *Salicineæ* from *Salix; Berberideæ* from *Berberis; Tamariscineæ* from *Tamarix.*

2. When the genus from whence the name of the family is derived has a name of inconvenient length, and there is not in the family a tribal name formed from the same generic name, the termination *eæ* is admitted; as *Dipterocarpeæ* from *Dipterocarpus.*

3. For some very large families universally known under their exceptional names, the ancient designation is preserved, as *Cruciferæ, Compositæ, Gramineæ,* etc.

4. An old generic name no longer preserving that rank, but applied only to a section or even a species, may be maintained as the base of a family name, as *Hippocastaneæ* from *Æsculus hippocastanum.* (DC.)

ZOOLOGICAL FAMILIES, SUBFAMILIES AND TRIBES.

§ **XVII.** The names of zoological families are best formed by adding the termination *idæ* to the name of the earliest known or most characteristic genus contained in them; and of subfamilies by adding the termination *inæ;* this should be done by changing the last syllable of the genitive case into *idæ* or *inæ.* Ex. *Strix, Strigis, Strigidæ;* not *Strixidæ; Buceros, Bucerotis, Bucerotidæ;* not *Bucerosidæ,* or *Buceridæ.* (DC., B. A., R., Bourg., D'Orbigny, etc.)

This practice is strongly recommended in both editions of its rules by the British Association Committee. It has been adopted so generally by modern authors, and its simplicity and convenience are so great, that it is not necessary to enlarge on its claims for consideration. Thorell suggests *oidæ* as a termination, which is certainly less euphonious, if in some cases more classical. Scudder and a few others would attempt to retain old forms, however diverse, a course apparently both impracticable and undesirable.

There are a few generic names, which, it is claimed, will not readily receive these terminations, and some objections have been made to insisting on the universal application of this terminology.

The desirability of uniformity and the self-evident convenience of being able to recognize the value of a group at a glance are weighty reasons in favor of the practice which, it appears to the Reporter, may safely be left to the common sense and good taste of naturalists, which in the course of time will practically decide the question; especially since no fundamental principle is endangered by allowing some latitude in the matter.

§ XVIII. Botanical subfamilies (*subordines*) are formed from the name of one of the genera contained in them with the termination *eæ* or *ineæ*, and also the names of tribes and subtribes which take the termination *eæ*. Ex. *Roseæ* from *Rosa*, etc.

Permanency of Names of the Higher Rank.

§ XIX. Names of higher rank than genera, with the fluctuation of their limits caused by the advance of science, are not rigidly subject to the *lex prioritatis*.

While this generalization has not been formally enunciated in the B. A. rules it has become practically the general usage of naturalists. Thorell explicitly adopts it, and indeed it is impracticable to follow any other course, especially in relation to the more ancient names. A time will doubtless arrive when mutations in the names of the higher groups, particularly families, will be as unnecessary as they are undesirable, but in zoology that period has not yet come.

It should be clearly borne in mind that such changes are only allowable when by mutation of the characters or through newly discovered facts, the name in question has become glaringly erroneous, or liable to introduce errors or confusion into science. In family names this occurs most often when a genus from whose name that of the family may have been taken is removed from association with the majority of genera which that family has included, and the said genus is inserted in another family which has already a well established name. Also, when a large number of genera are re-distributed into families, widely differing in their limits from those in which they had previously been known. In either of these cases the liability to induce error may be so great as to render a new name desirable. The answers to query XXIII of the circular indicate that a majority of American naturalists concur in this conclusion.

On names of Genera and Subdivisions of Genera of Higher Rank than Species.

§ XX. Genera, subgenera, or sections, receive names, preferably substantives, which bear the same relation to each other that the proper family or surnames of individual men do to one another. (DC., Bourg.)

These names may be taken from any source whatever, or may be framed in an absolutely arbitrary manner, subject to the following conditions. (DC., G. R. G.)

De Candolle justly remarks that it is with generic names as with our patronymics. Many surnames are inconvenient or even absurd from bearing an adjective form, from having an inapplicable meaning, on account of being difficult to pronounce, or for some other reason. But, since they actually exist, why should they be changed? It is not the end of science to make names. She avails herself of them to distinguish things. If a name is properly formed and different from other names the essential points are attained.

Generic names may be taken from certain characters or appearances of the group, from the chief habitat, names of persons, common names and even from arbitrary combinations of letters. It is enough if they are properly constructed and do not lead to confusion or error.

As long as these matters are not judged by this most important principle, so long must nomenclature suffer from the proposition of rules admitted by some and rejected by others. (DC.)

Differing names are sometimes formed in honor of the same person when their names lend themselves to it. *Pittonia* and *Tournefortia* are derived from Pitton de Tournefort; *Brownia* and *Brunonia* from Brown, etc. These names should be preserved for they cannot be confounded in speech or in the tables. Certainly, if after *Brownia* had been proposed there should appear a naturalist named *Brunon*, no one would criticise a genus *Brunonia;* hence *Brunonia* is an admissible generic name. (DC.)

The B. A. committee strongly object to names arbitrarily formed and having no meaning. On the other hand, if euphonious and constructed according to the Latin form, they are not without their advantages. Yet it is certain that names having a definite meaning are much to be preferred.

§ XXI. Minor subdivisions of a genus may be indicated by a name, or, if of less value than a subgenus, may preferably bear merely a letter or a number without a name. (DC., etc.)

§ XXII. When the name of a genus or of one of its subdivisions is derived from the name of a person, it should be constructed in the following manner. (Bronn, Bourg., Th.)

A. The name disembarrassed from all titles and all preliminary particles is terminated by *a* appended to the form of the genitive case, thus taking on a feminine form.

The following examples illustrate the method both for generic and specific names.

Name	Brun	Bruni	Bruno	Brunus.
Genitive	Bruni	Brunii	Brunoi	Brunusi.
Generic form.	Brunia	Bruniia	Brunoia	Brunusia.
Adjective form.	Brunianus	Bruniianus	Brunoianus	Brunusianus.

Name.	Bruna	Brune	Bruny.	
Genitive	Brunæ	Bruni	Brunyi or Brunii	
Generic form	Brunæa	Brunia	Brunyia or Bruniia	
Adjective form.	Brunæ	Brunianus	Brunyianus or Bruniianus.	

B. The syllables which are not modified by this termination preserve exactly their original orthography, even to the letters or diphthongs employed in certain languages, but which are not used in the Latin tongue. However, the *ä, ö, ü,* of Germanic languages become *æ, œ, o;* the é and è of the French language become *e; y* at the end of a word of one syllable is treated as a consonant *(Quoy, Quoyia; Gay, Gayia*), and mute *e* final becomes *i*, or is dropped entirely (*Perouse, Perousia*).

To this proposition the B. A. committee agree, and it is generally followed.

C. The genitive form is used for specific names when the proper name is that of the person who collected or originally described the species, the adjective form is proper in all other cases. Thus *Corvus corax,* Brun non Linnæus, or a new *Corvus* collected by Brun, would be *C. Bruni.* A *Corvus* named after one's friend Brun, or an ornithologist Brun, would be *C. Brunianus.* (Bourg.)

An adjective form, however, must in no case be given to a generic or subgeneric name; e. g., *Wolfartaria* Gray, named after Mr. Wolfart should be *Wolfartia.* (B. A., Bourg.)

§ XXIII. It is highly undesirable that genera should be dedicated to persons not in some way connected with the study of the natural sciences, or with the collection of materials upon which that study is based. (B. A., Bourg.)

The recommendations of the B. A. committee contain an objection to personal names in zoological genera, but the practice has been carried to such an extent that even if it were still less desirable, any prohibitory rule or recommendation would be unavailing. However, names having a modern political or religious significance should be particularly avoided.

§ XXIV. Naturalists who may propose generic names will conform to good taste and benefit nomenclature by regarding the subjoined recommendations.

A. Avoid very long names and such as are difficult to pronounce. As a general rule it may be recommended to avoid introducing words of more than five syllables. Ex. *Craxirex; Eschscholtzia; Thecodontosaurus; Strongylocentrotus*, etc. (DC., B. A., etc.)

B. Indicate the etymology of each name.

Great confusion has resulted from endeavors on the part of authors to correct names of which the etymology was uncertain, but which seemed to them to be erroneously constructed. When the etymology is given this question can be accurately and promptly decided and no confusion result. (B. A., DC., etc.)

C. If an author has proposed a genus which has not been admitted, he should most carefully avoid creating another under the same name.

Nothing is more inconvenient in synonymy than to have to explain that a certain genus of any author is not the genus bearing the same name and of the same author, but of a different epoch. In general, a generic name which is known to have once been used in botany or zoology, should be studiously avoided forever afterward in any other connection in the department in which it was once introduced. (DC., B. A., etc.) See under head of names to be rejected.

D. Avoid forming names from barbarous or savage tongues unless they may be found more or less frequently cited in books of travel, and present a euphonic combination easily adapted to the Latin forms. (DC.)

E. Recall, if possible, in the composition or termination of the name the affinities or analogies of the genus. Ex. Many genera of fossils appropriately end in *ites*, many of ferns in *pteris*, etc.

F. Avoid all adjective names.

The names of genera are in all cases essentially substantive, and hence adjective terms cannot be employed without violence to grammatical construction, and their use is *prima facie* evidence of a want of good taste. The same may be said of names in the genitive case which are wholly inadmissible without reformation. (B. A., DC., Ver.)

G. Avoid giving names which in sound or spelling closely resemble other names already proposed, even when the etymology of the two words is diverse. Ex. *Leuco-dore, Leucodora; Otostomia, Odostomia*, etc., etc.

The danger of confusion in such cases is self-evident, and the naturalist mindful of the objects of nomenclature will endeavor to avoid them.

H. In compounding a name from two other names, the essential or radical parts of both should be retained and the changes confined to their variable terminations. (B. A., DC., Th.)

A name compounded of the first half of one word, and the latter half of another, is an ungrammatical monstrosity; e. g., *Loxigilla*, from *Loxia* and *Fringilla*.

In other cases when the commencement of both words is retained, the excision may still be too great; e. g., *Bucorvus* from *Buceros* and *Corvus*.

In general compound words are open to objection from their too great length, or the liability of introducing a barbarism in endeavoring to render them shorter; but, when compounded with care and not too long they may occasionally be used with advantage in designating intermediate genera. (B. A.)

I. Avoid compounding names from two different languages. (B. A., Th.)

Such names are great deformities in nomenclature, and are, when one of the parts is of an adjective character, reasonably considered by most authors as subject to reformation or rejection.

J. Geographical names, being for the most part adjectives, should be avoided in naming genera. (B. A.) The termination *ensis* is only applicable to names of species derived from the name of their habitat.

K. Technical names expressing trades or professions should only be used when they have a special relation to the habits or characters of the organism designated. (B. A.) Ex. *Arvicola; Pastor; Regulus*, etc. If this caution were disregarded, such names would be highly objectionable.

L. Mythological names are best applied in cases where a direct allusion can be traced between the narrated actions of a personage and the observed habits or structure of an animal. Thus when the name *Progne* is given to a swallow, *Clotho* to a spider, *Nestor* to a gray-headed parrot, etc., a pleasing and beneficial connection is established between classical literature and natural science. (B. A.)

M. Names expressing positive characters are to be preferred to comparative names in most cases, though occasionally a diminutive may be employed without disadvantage.

§XXV. In naming subgenera or sections of a genus beside the recommendations under the last head, the following additional precautions may advantageously be attended to. (DC.)

A. For the principal divisions of a genus, if it be decided to apply names to them at all, it is well to adopt such as recall the genus itself by some modification or addition.

B. The chief modifications in use in nomenclature are as follows:—before a Greek derivative *Eu* and *pseudo;* after it, *astrum, oides,* or *opsis;* before a Latin derivative *sub;* after it *ella, una, ina, ites,* etc. Usage has justified to some extent the application of these modifications to words of uncertain etymology or arbitrary formation, in connection with which the Greek syllables are best entirely avoided.

C. The application of these modifications should be governed by the subjoined restrictions.

(1) The prefixes *pseudo* and *sub* should, especially the latter, be reserved for use with specific names, though the former may rarely be applicable to the name of a generic group which has been confounded with another whose name is not of Latin extraction. *Eu* may be used before generic names when they are derived from the Greek language.

So far as specific appellations are concerned *pseudo* may be employed when it is desired to connect the name of a species with that of another with which the former has been confounded. (Bourg.)

Sub may be used in designating a new species before the name of another with which the first has intimate relations. It has also a few legitimate Latin compounds which may be used for specific names, such as *subterranea*, etc.

(2) The suffixes *ella, una, ina, ites* (Latin), and *astrum, oides, opsis* (Greek), etc., may be applied to generic or specific names.

The termination *ella, una, ina,* are used at the end of generic names derived from the Latin, to indicate that the genus indicated by the radical to which the termination is appended, resembles, in some way, the new genus.

They are also used in reforming a name which is inadmissible for any reason, in order to preserve a convenient similarity. For instance, *Cæcilia,* if employed for a shell (but already in use in vertebrates or insects) might be modified to *Cæcilianella,* in order that convenience in the consultation of tables or indices might be conserved for the new name in connection with the old one.

In the first sense these suffixes are also used in forming specific names.

The termination *ites* is found convenient for designating fossil organisms analogous to the living form, whose generic name is the radical to which *ites* is appended. It is rarely used with specific names. The terminations *astrum, opsis, iscus,* etc., may be used in connection with generic or specific names.

Bourguignat suggests that *astrum* be appended to the name of a genus to indicate its typical subdivision, and be reserved for this purpose, but this has not been adopted.

A subdivision of a genus should not be formed by adding to the name of the genus as a radical, the termination *opsis* or *oides*, but on the contrary these terminations should be reserved for subdivisions, which resemble another genus by adding them to the name of that other genus, when it is of Greek origin. (DC.) Since a subgenus necessarily recalls or resembles its genus the announcement of the fact in its name is a work of supererogation. To annul such names which have been already imposed, however, would be still more inconvenient in spite of their objectionable character.

The termination *oides* is inelegant when applied to generic names or those of higher rank, and is better reserved for specific names of Greek or barbarous (never Latin) origin, and used in the following cases, only, if at all.

a. When the radical of the specific name is the name of a genus which it resembles : e. g., *Helix naticoides* for a species resembling a *Natica*.

b. When the radical is the root of the name of another species which the new one resembles; e. g., *Helix carascaloides*, a species like *H. carascalensis*, or *Unio Moquino- ides* for a species like *Unio Moquini*.

One single apparent exception to the compounding of Greek and Latin has been consecrated by usage; namely, *ovoides* from *ovum* and εἶδος; this has arisen from the absence of euphony in the correct form, *ooides* (ᾠόν and εἶδος). (Bourg.)

D. For subsections or groups of species, the name of the species around which the others may naturally be clustered, may be used with the termination *iana*, as a group of species of *Helix* related to *H. pomatia* may be indicated by the term *Pomatiana*. (Bourg.)

E. Avoid, in designating sections, names which have previously been employed for genera or sections of other genera.

Sections are chiefly used in botany, being seldom named in zoological literature. Even in botany they are not usually cited in synonymy, and hence it has happened that the name of a section has been permitted to exist simultaneously with an identical word applied to a genus, or the section of another genus.

This, however, should be avoided, especially as at some later day, the sections may be advanced to subgeneric or generic rank.

F. When the name of a section is cited it is always announced in conjunction with the generic and specific names, and is placed between them in parentheses. (DC.)

G. In defining new genera or sub-genera, a typical species should always be mentioned to serve as a standard of reference. (B. A.)

Of Specific Names and Names of Subdivisions of Species.

OF THEIR FORMATION.

§ XXVI. Each species, even when it is the only one of its genus, is designated by the name of the genus to which it belongs followed by a name called specific, which consists of a single simple or compound word, and has usually the character of an adjective. (DC., Bourg.)

See also § XXII. A. In certain exceptional cases, such as gall insects, the connection by a hyphen of a third word, with the specific name proper, when justified by usage, is not considered as an infraction of this rule. (Riley.)

§ XXVII. The specific name should, in general, indicate some feature of the appearance, characters, origin, history or properties of the species.

If it is taken from the name of a person it is usually for the purpose of recalling the individual who described, discovered or was in some way connected with it, or with the study of the group of which it forms a part. (DC., Bourg.)

If a mythological personage has furnished a name which has been applied to the genus, it is permissible for the species to receive the names of his family or tribe. A series of names having a mental association with one another may be used for the species of a genus, as for instance, when a group of wood-butterflies receive the names of Indian chiefs, native to the region inhabited by those insects; or the names of various mythological heroes or warriors are associated with the respective species of a group characterized by martial colors, habits or bearing. These names, unless already euphonious, should receive a modified Latin termination as already indicated.

§ XXVIII. A specific name may be a proper noun or substantive, in which case it does not necessarily accord in gender with the generic name; but when the specific name has the character of an adjective, its termination should be constructed to agree in gender with the generic name.

In the first case it is frequently written beginning with a capital letter; in the second case, it should commence with a small letter. It is probable that in all cases it would be better to commence the generic name with a capital, and the specific name with a small letter, except possibly when it is derived from the name of a person.

On this point, however, naturalists are divided in opinion, usually following the rules of their respective vernaculars.

§ XXIX. The same specific name must not be used for two species of the same genus, but, to species of different genera, the same specific name may be simultaneously applied.

§ XXX. In constructing specific names naturalists will do well to regard the following recommendations. (DC., Bourg., B. A., etc.)

A. Avoid names of too great length or difficult to pronounce.

B. Avoid names which express characters common to all or nearly all the species of the genus.

Such names were formerly rejected, but the confusion thus caused in nomenclature was far worse than the want of definitiveness in the names changed.

C. Avoid geographical names taken from little known localities or of narrowly restricted application, unless the habitat of the species is known to be equally restricted.

D. When a geographical name is applied to a species if there be a classical form of it, it is to be used in preference to the modern name (e. g., Anglia, not England) unless the Latin name be extremely obscure, inharmonious, or almost unknown.

Names not represented in classical literature preserve the radical intact, latinizing only the termination.

The name of a locality is always terminated by the suffix *ensis* (e. g., a species from Carascal would be *carascalensis*).

The name of a river or other body of water, a province, a country, or a kingdom, on the contrary, is terminated by *ius, icus, inus, itus,* etc., if the gender be masculine; or by *ia, ica, ina, ita,* etc., if it be feminine; as *arabicus, euphraticus, algirus* (if the province of Algiers) *appennina, anglica, cypria, texasiana, algiriensis* (if the town of Algiers), etc., etc.

Since geographical names are often liable to change from the variation of political boundaries, they are in many cases particularly objectionable. Formerly some authors rejected them entirely, or tolerated them only when exclusively applicable. Such mutations are no longer approved by the majority of naturalists, since fixity is the fundamental principle to be regarded in nomenclature, yet as these names are liable to objection on serious grounds it is desirable to make as little use of them as possible. (See names to be rejected.)

E. Avoid using in the same genus names too closely similar in form or sense, above all those which only differ in their last letters.

A judicious naturalist, for example, will not call a species *virens* or *virescens* in a genus already containing a species named *viridis*. (B. A.)

F. Manuscript names found in the notes of a collector or on his herbarium tickets, when they are well selected, may be adopted, but not without supplying a description and stating that the name has not elsewhere been formally introduced into nomenclature.

Without these restrictions or precautions the use of MSS. names is highly objectionable, and has already been the cause of great confusion and annoyance to naturalists. The MSS. names of Beck, Solander, Leach and others, have long been stumbling blocks in the path of science from having been quoted by naturalists with no reference to the fact that they are to this day undescribed, and therefore wholly valueless.

G. Avoid names which have already been employed in the genus or a related genus, and which have fallen into synonymy. (DC.)

H. Do not dedicate a species to a person who has not discovered, described, figured nor studied in any manner, it or the group to which it belongs.

I. Avoid names similar to names employed to designate the genus to which names belong or related genera.

§ XXXI. Names of subspecies or varieties are formed like specific names, and, when cited, follow the name of the species to which they are related, preceded by the abbreviation *subsp.* or *var.* They should never have a substantive character.

The practice which has lately arisen of writing these names without the abbreviation is objectionable, as incompatible with binomial nomenclature.

Of Names of Hybrids, Sports and Cultivated Varieties used in Botany.

§ XXXII. Hybrids whose origin is certain are designated by the name of the genus followed by a combination of the specific names from which the hybrid has proceeded. The name of the mother species is placed first, terminated by an *i* or an *o* and connected by a hyphen with the name of the species which has furnished the pollen; e. g., *Amaryllis vittato-reginœ*.

Hybrids of doubtful origin are named like species, and are preceded by the sign × placed before the generic name; as × *Salix capreola*, Kern.

§ XXXIII. Subvarieties, variations and subvariations of wild plants may receive names analogous to those of varieties, or simply numbers or letters which may facilitate their classification.

§ XXXIV. Seedlings, or sports, from cultivated plants receive fancy names in the vernacular which should differ as much as possible from the Latin designations of species or varieties. When it is desired to cite them with botanical species, subspecies or varieties, they are indicated by the succession of the names: as "*Pelargonium zonale* Mistress Pollock." (DC.)

The above limitations, as explained by De Candolle, are necessary to prevent confusion between the principal natural modifications of a species and the illimitable number of seedlings, sports, and inferior modifications of cultivated plants, the product of gardens; whose transitory appellations if proposed in the Latin form, would (as some have already) creep into the literature of the subject, be confounded daily with species or varieties, and produce unending complications.

Of the publication of Names and the date to be assigned to each Name or Combination of Names.

§ XXXV. The date of a name or combination of names is that of their actual publication, that is to say from the time of occurrence of such circumstances as give to the name an irrevocable publicity.

§ XXXVI. Publication consists in the public sale or distribution of printed books, pamphlets or plates. To this category botanists add the distribution or sale of specimens of the plants described, to which written or printed tickets containing the proposed new names are attached. (DC.)

To constitute publication nothing less than the insertion of a distinct exposition of essential characters in a printed book can be deemed sufficient. (B. A.)

Satisfactory plates or figures which express the essential characters of the organism concerned, and on which the proposed name is engraved or printed, are generally held to be equivalent to a definition.

The exception in favor of published herbaria by botanists, appears injudicious, and is the result of an ancient usage which fortunately has never obtained in zoology, and as a method of publication in modern days can be seldom necessary and never advisable.

§ XXXVII. A communication in a public assembly, of names applied to specimens in collections, the ticketing of plants in public gardens, or the reading of a paper containing new names before a scientific society, does not constitute publication. (DC., Th., Benth., V., B. A.)

Communications to a public assembly unless published in an immediate literal report of the meeting, rest only on the uncertain recollections of the audience. Labels in public museums and gardens may be transposed or replaced from hour to hour. Communications to a society may be subject to amplification, correction or revision, up to the time of printing. In all the above cases the *fact* of publication cannot be placed above question.

The inadmissibility of adopting the date of reading of a paper is shown by supposing a case in which it failed altogether to be printed. No one would then assert that it had actually been published. Yet that such honor or credit as may be due to the original discoverer of a fact, can be, and is, secured to him by such an announcement of it, whether he be the first to print it or not, no one will deny. This seems to be the principle contended for by the few who demand priority from the date of reading. It is possible that some believe that credit or honor is secured by the application of new names, but scientific nomenclature owing its right to exist to the fundamental principle of fixity, should not be allowed to suffer from considerations of this nature.

That the majority of American naturalists concur in the above, is manifest from the replies to query V of the circular before alluded to.

§ XXXVIII. The date borne by a publication is presumed to be accurate until the contrary is proved. (DC.)

It is, nevertheless, well known in some cases this date is not correct, or represents the presentation instead of the publication of the matter bearing it. This is, for instance, the case with several of D'Orbigny's works and with the transactions of some learned societies.

Such absence of frankness and plain dealing in matters where the utmost precision is desirable, is highly reprehensible. It would hardly be unfair in instances where a doubt arises as to priority between an honestly dated work and one issued as above, to throw the *onus probandi* on the publishers of the latter.

§ XXXIX. A species is not to be considered as named unless both generic and specific names are simultaneously applied to it. (DC.)

This follows from § XXVI, and the majority of the replies to query XVII of the circular, concnr in it.

§ XL. A species announced in a publication with both generic and specific names, but without any description, cannot be considered as published. It is the same with a genus simply announced by name. If, subsequently, the author or some one else, makes public what this name signifies, the date of the second publication is the only one to be taken account of. (DC.)

A specific name without a generic name, or a combination of generic and specific names without a description of any kind, is nothing. They are empty words without sense. They only acquire a value from the time when some one gives them sense by completing them.

It may be said that there are specific descriptions so short or badly made as to signify almost nothing, aud hence these should be disregarded as incompletely published, or names pure and simple should be admitted. There is, however, a difference between the two cases. The absence of all characterization of a name is a definite, positive fact. The insufficiency of a description is something vague, which may be contested. Besides, an apparently insignificant word may sometimes lead to the identification of a species. (DC.)

In revising a group in which such names occur, the naturalist who desires to avoid multiplication of synonyms, may choose to adopt the unpublished name; a course which is to be commended only when by some particular circumstances the absolute identity of the group about to be named, with that intended by the author of the unpublished name, can be established. (G. R. G., etc.)

§ XLI. A generic name accompanied by a description, but without a reference to it of any definite species, may be considered as published when the organisms to which it is intended to refer, are unmistakable.

Compare circular, XVI.

§ XLII. A generic name unaccompanied by a description, but under which, as type, a single new species is fully described or satisfactorily figured, and concerning which there can be no uncertainty, may be admitted as published. Conversely, when a generic name is fully characterized, and a new specific name without a description is appended to it as the type of the new genus, and about which no doubt can be entertained, the species may be regarded as published. (Th.)

The practices mentioned are doubtless highly objectionable. But in the first case the generic characters may be regarded as included in the specific diagnosis, and in the second the specific characters may be considered as expressed in the generic description; in either instance a doubt might reasonably exist as to which of all the characters of a new organism were to be regarded as generic, and which as appertaining to the species.

This differs from the cases in which a generic name without a diagnosis is placed before the names of a number of species then or previously described. In these instances, if there are characters warranting the allocation of species, and separating them from other genera, it is the duty of the propounder of the genus to indicate them to the best of his ability. If too ignorant or too indolent to attempt the task of differentiation in words, his work may safely be regarded as unworthy of recognition. (B. A., 11.)

§ XLIII. Naturalists will do well to bear in mind the following suggestions :

A. Indicate exactly the date of publication of each work or portion of a work, or of the distribution of named specimens of plants under § XXXVI. (DC.)

It would be very useful if scientific journals or works on bibliography would denote the exact date of reception or actual publication of volumes or plates appearing successively, or in which one is liable to be deceived by the title pages, or where doubt arises from the absence of dated title pages. This is especially the case with works appearing in parts. In regard to plants in well conducted herbaria, the date of reception (which is almost always that of distribution) is usually noted on the labels. (DC.)

B. Do not publish a name without clearly indicating the nature of the group (tribe, genus, section, species or variety) to which it is supposed to belong. (DC.)

C. Avoid mentioning or including in any publication unaccepted manuscript names, above all when the persons who gave these names have not formally authorized their publication. (DC.)

To publish a name which cannot be adopted, or which is undefined by the person citing it, or some previous author, is wantonly to throw a synonym into circulation, at least in tables and bibliographies.

Our nomenclators would be of double size if they contained all the names which exist in travellers' note books, museum and herbaria, and which have not the slightest value. Names of this kind when printed are still-born. Why then augment their number unless at least the author asks that they be made public.

Of the Precision to be given to Names by the Citation of the Author who First Published them.

XLIV. To be exact and complete in the indication of the name or names of any group, whatever, it is necessary to cite the author who has first published the name or combination of names in question.

A. When it is desired to indicate that the author proposed the whole name in the form in which it is cited, his name or an abbreviation of it simply follows the specific name; as *Anomia craniolaris*, Linnæus, or Lin. or L.

B. When it is desired to indicate that the application to the species of the specific name is all that is due to the author cited, the abbreviation of his name may be either (1) enclosed in parenthesis [e. g., *Crania craniolaris* (L.)]; or (2) followed by the abbreviation *sp.*, as *Crania craniolaris* L. sp.

C. When it is desired to indicate the author of the combination cited, together with that of the person who first proposed the specific name, the name of the latter or its abbreviation is usually written in parenthesis followed by the name of the author of the combination cited (e. g., *Crania craniolaris* (L.) Nilss.); or the name of the author of the combination follows the specific name separated from that of the author of the specific name by the word *ex* (example, *Crania craniolaris* Nilsson ex L.). The latter mode does away with parentheses and seems preferable.

D. When it is desired to indicate still more fully the relation of the author of the specific name to a former generic combination, the name of the author of the cited combination is followed by a parenthetical clause consisting of the name of the author of the specific appellation separated

from the designation of the genus to which he referred it by the word
sub ; as *Crania craniolaris* Nilsson (L. sub *Anomiœ*). This form is chiefly
employed in Botany, and is liable to objection from its length. (cf. Lec.
p. 204, et seq.)

For many years, Linnæus, Lamarck, Cuvier, Blainville, D'Orbigny and the majority
of zoologists, as well as the universality of botanists, in citing a combination of words
forming a specific name, appended to it as *authority,* the name of the author of the
combination, whether he were the author of the specific name or not.

More lately some zoologists have followed a different method, recommended by
Strickland in the B. A. rules of 1842, but which was from the first strongly combated by
Agassiz, and other influential naturalists.

This method consists in appending the name of the author of the specific name in
all cases without reference to the generic combination.

In deference to the said rules it has been somewhat extensively practiced, though
open to several objections which have been exhaustively set forth by Agassiz, D'Orbigny,
Bourguignat, and De Candolle; the latter remarking that, if followed, it would cause
zoological lists to resemble a city directory in which the names of individuals were
classed according to the alphabetical order of their Christian names.

Numerous battles of words have taken place over the questions which is, at most,
not one of fundamental importance, and it appears to your Reporter that it may safely
be left to the common sense of naturalists to decide.

There can be no question that the advantage of precision is with the older method,
which accords with the practice of Linnæus and the spirit of the Linnæan canons,
while, by the mode suggested in ¶ C, no uncertainty is permitted, and the most tender
sentiment of justice, toward both the describer and the referrer of the species, is satis-
fied.

§ XLV. In writing the name of a species with the authority, it is unnecessary to separate the latter from the specific name, by a comma. (V.)

There has been little uniformity in regard to this matter, and Prof. Verrill has judi-
ciously suggested that "the best usage appears to be without any punctuation, the
authority in this case being understood to be a noun in the genitive case, though written
in the nominative form or more frequently abbreviated."

§ XLVI. A change of the diagnostic characters or the circumscription
of a group does not authorize the citation of another author than the one
who first published the name or combination of names.

When the changes have been considerable, to the citation of the primitive author may be added *mutatis char., pro parte, excl. gen., excl. sp., excl.
var.,* or such other abridged indication as may indicate the nature of the
changes in the revised group, or of the group treated of. (DC.)

Whatever the changes brought about by the advance of science, in any group the
fact that a certain author proposed a certain name is the one positive incident which
can be recognized. (DC.)

Some authors (Riley) object to this generalization, but it appears to be generally
accepted.

§ XLVII. Names published from manuscripts, unpublished catalogues,
museum labels and herbaria, etc., are given precision by citing as authority the author who first published them in spite of any indication to the
contrary which he may have given. Similarly names used in botanical
gardens are cited as of the author who first published them. In the full

text or synonymy, the manuscript, herbarium, museum, collection or garden, should be cited. (*Lam. ex Commers. in herb. Paris: Lindley ex horto Lobb; Martyn ex coll. Soland.; Gray ex Leach* MSS., etc.) DC.

The publication of the name is the essential fact, for it is that which preserves the name from change except for grave reasons. The publisher has taken the principal step. The traveller who has obtained a specimen, and has perhaps given it a provisional name in his collection, merits our gratitude, and may have more claims to it than the editor of the name; whence it becomes extremely proper to cite him, in the mention of the habitat or the collection accompanying the description. But it is not he who has made it public at a certain date, and if he had been consulted perhaps he would have applied another name to it. It is in the interest of precision therefore to cite, in botany, for example, the name of Spruce for a plant named and distributed by him, though described later by Bentham; and to cite Bentham for a plant of Hartweg, distributed by the latter under a number without a name, but afterward named by Bentham. It would be inexact to act otherwise, and unjust to the old naturalists. Commerson, for example, left plants in herbaria without publishing them. If they were published now, so great have been the changes in the science of Botany, they would indicate a condition of things of which that zealous collector, were he living, would be probably among the first to re ognize the falsity. (DC.)

This, however, cannot be held to apply to cases where one working naturalist submits to another the result of his studies, to enable the former to complete a monograph or correctly interpret a fauna. In such instances the work may reasonably be considered a joint affair, and the names due to the one or the other as the case may be.

§ XLVIII. When an existing name is changed in value, from being raised to a higher rank or reduced to a lower rank than that it originally held, the change is equivalent to the creation of a new group, and the author to be cited is he who has made the change. (DC.)

It is a common but objectionable practice to cite the author of a section or family as the authority for a genus or order, or the reverse. By this negligence the opinion of the primitive author is misrepresented, and the reader is deceived in regard to the date of later name. (DC.)

§ XLIX. The authority following the name of an organism is unless very short, usually indicated by an abbreviation. To effect this the particles or preliminary letters which do not strictly form part of the name are dispensed with, then the first letters are indicated without omitting any. If a name of one syllable is sufficiently complicated to be worth the trouble of abbreviating, the first consonants are indicated (Br. for Brown); if the name has two or more syllables, the first syllable plus the first letter of the second (or the first two if they are consonants) is made use of (Juss. for Jussieu; Rich. for Richard).

When less abbreviation is desirable to avoid confounding two names commencing with similar syllables, the same system is followed by giving the first two syllables with one or two of the consonants of the third, or adding one of the last characteristic consonants of the name (Bertol. for Bertoloni, to distinguish it from Bertero; Michx. for Michaux as opposed to Micheli; or Lamx. for Lamouroux as distinguished from Lamarck).

Christian names are abbreviated in a similar manner (Adr. Juss. for Adrien Jussieu; Gaertn. fils or Gaertn. f. for Gaertner junior, etc.).

When a usage is established for an old name in some other way it is better to follow it, as L. for Linnæus; St. Hil. for Saint Hilaire. (DC.)

This rule has been followed by Linnæus, De Candolle, and all botanists up to a recent date. No particular rule has been laid down for the guidance of zoologists, but this will doubtless recommend itself to them when considered. In some later works, only those familiar with the literature can divine whether *Bth.* is the equivalent of Bentham, Benth or Booth, *Sz.* of Schultz, Steetz or Szowitz; or what is the equivalent of *Htsch., Hk., H.Bn., Bn., Btt., Lm., Reich.* or *Spng.*

Of Names to be preserved in writing, Dividing or Modifying the Limits of Existing Groups.

§ L. A change in the diagnostic characters, or a revision which carries with it the exclusion of certain elements of a group, or the inclusion of new elements, does not authorize the change of the name or names of a group.

§ LI. When a group or genus is divided into two or more groups, the original name must be preserved and given to one of the principal divisions. The division including the typical species of the primitive genus, if any type had been specified, or the oldest, best known or most characteristic of the species originally included when the primitive genus was first described by its author, is the portion for which the original name is to be preserved. If there is no section specially so distinguished, that which retains the larger number of species should retain the old name (DC.), but the latter cannot be applied to a restricted group containing none of the species referred to the primitive group by its author at the time when it was described or when he enumerated the species contained in it.

The majority of the replies to query XII of the circular concur in the above.
According to Linnæus the name should remain with the most common and officinal species; an equivocal expression if there is one which is most common and another the officinal species. The *Convolvulus sepium* and the *Erica vulgaris* were very common and very anciently named species when Brown made of one the genus *Calystegia*, and De Candolle of the other, his genus *Calluna*. It was, however, much better to do this than to change the names of a hundred species of *Convolvulus*, and two hundred of *Erica*. When there is no authoritative type, the number of species should always be taken into consideration. (DC.)

§ LII. When an author has specified no type, it is then necessary, in dividing his genus, to retain his name for the subdivision containing the species, which the next subsequent author, treating of the genus has specified or regarded as the typical exemplar. B. A. If no subsequent author has selected a type, the first species of the primitive author may frequently be taken as the type, or a species may be selected from among those originally specified as belonging to the genus when it was formed, due regard being paid to the necessity of retaining as many of the original species as possible in the division which is to retain the old name.

It would, manifestly, be liable to introduce errors and confusion, if it were insisted that the first species should invariably be taken as the type, or were it permitted to take species subsequently added to the group, and which the original author did not know when he established his genus. No arbitrary rule will suffice to determine, off-hand, questions of so much complication as is often the decision in regard to the type of an ancient genus which has been studied by a number of authors.

In the first of the above cases lists are often arranged in alphabetical or faunistic order, or the aberrant species are placed at or near the beginning and end of the list, while the more generalized and characteristic species are put between the others. In the second case aberrant species might be added, and subsequently taken away from the genus, carrying with them the name consecrated by the primitive author to the very group which the subsequent reviser might then seize on for his own. Still more the aberrant species carrying the primitive generic name might subsequently be found to belong to a genus described before the one revised. Then the name originally given to a valid group might be subject to rejection as a synonym, while the valid group itself which originally bore that name was rejoicing under a new appellation received from the industrious revisers! Absurd as it may appear, mutations similar to this might be mentioned.

The answers received to questions on this point in the circular, will be seen to be by a large majority in concurrence with this section.

§ LIII. In dividing a genus of which there are already synonyms, if these synonyms or any of them are typified by the same species or group of species as that or those originally selected as types for the primitive genus, the names should be cancelled *in toto*, and not used for the re-stricted subdivisions. (B. A.)

To use strictly equivalent synonyms, in a new sense for different divisions in one family, is sure to create confusion and necessitate lengthy discriminating passages in subsequent synonymical work. When the so-called synonyms are founded on species belonging to different sections of the genus, although the names may have been con-sidered as co-extensive in their application, it is desirable to use these names to indi-cate the divisions of the genus when it may be revised. (B. A.) In fact there is hardly any difference between the latter case and the revival of a valid but forgotten name for the group properly designated by it, and to which another legal name cannot be applied.

§ LIV. In the case of the consolidation of two or more groups of the same nature the oldest name must be retained for the whole. If both or all are of the same date, the reviser may select the one to be retained. (B. A., DC.)

If a name of a genus be so defined as to be equal in extent to two or more previ-ously published genera it must be cancelled *in toto*. (B. A.) Example, *Tritonium* Müller, was so defined as to be equal to *Buccinum*, *Strombus* and *Murex* of Linnæus. Hence it should be wholly rejected. *Psaracolius* Wagler, is equivalent to five or six previously published genera, and must, therefore, be cancelled. (B. A.)

It follows from the above, that when it is necessary to unite several groups already named, the earliest unobjectionable name must be retained for the consolidated group, with a modified diagnosis.

§ LV. When it is necessary to divide a nominal species into several species, the form which first received the old specific name is the one which should retain it. (D. C., etc.)

§ LVI. When a section of a genus or a species is transferred to an-other genus in which a variety or other subdivision of a genus or species is already called by the same name as the group transferred, the group of

higher rank (as the species) retains the name while that of less value (as
the variety) takes a new appellation. (DC.)   Compare § LVII.

§ LVII. When a nominal genus is reduced to the rank of a subgenus or
a group nominally of subgeneric value is erected into a genus; when a
species becomes a subspecies or vice versa; their respective names remain unchanged, provided there does not result from it two genera or
subgenera of the same name in the animal or vegetable kingdom, two
species of the same name in one genus, or two subdivisions of a species
of the same name in the same species.   (DC.)

But in combining two groups previously considered as of equal rank, but of which
thereafter one is to be subordinate to the other, the newest (or least widely known
when both are of the same date) is the one to take subordinate rank; on the principle
set forth in § LIV.

Of a starting point for Binomial Nomenclature.

§ LVIII. The scientific study of different groups, having a value greater
than or equal to that of a class (classis), having been begun at different
epochs, and the inception of that study in each group respectively being
usually due to some "epoch-making" work, the students of each of the
respective groups as above limited may properly unite in adopting the
date of such work as the starting point in nomenclature for the particular
class to which it refers: *Provided*, — (1), that specific names shall in no
case antedate the promulgation of the Linnæan rules (Philosophia botanica, 1751); that (2), until formal notice by publication, of the decision of
such associated specialists (in such manner as may be by them determined
upon), shall be decisively promulgated, the adoption of the epoch or
starting point recommended by the committee of the British Association in
1842, namely, the twelfth edition of the *Systema Naturæ* of Linnæus
(1766), shall be taken as the established epoch for all zoological nomenclature.   Lastly, that (3), when the determination of the epoch for any particular group, as above shall have been made, the decision shall be held
to affect that group alone, the British Association date holding good for
all other groups until the decision for each particular case shall have been
made by the Naturalists interested in it, upon its own merits.   (Cf. Lec.,
¶ 1, p. 203, et. seq)

Recommended by the Reporter for adoption.

The question at issue is one upon which naturalists are unfortunately much divided
in opinion, and the difficulties arising from this diversity are becoming more injurious
to science with each succeeding year.

The decision is, of course, of fundamental importance in many branches of zoology.
In order that the question may be better understood, a brief historical statement is
subjoined.

A series of rules for nomenclature was to some extent foreshadowed by Linnæus in
his *Fundamenta Entomologia* of 1736.   These rules were first definitively proposed in the
Philosophica Botanica, which appeared in 1751.   These rules, however, related almost
exclusively to the generic name or *nomen genericum*.   In 1745 he had employed for a
few species of plants for the first time a specific name (*nomen triviale*) composed of one

word, in contradistinction to the polynomial description of a species (*nomen specificum*) which was previously the rule among naturalists. That which now seems the most happy and important of the Linnæan ideas, the restriction of the specific name as now understood, seems to have been for a long time only an accessory matter to him, as the *nomina trivialia* are barely mentioned in his rules up to 1765.

In 1753, in the *Incrementa botanices*, while expatiating on the reforms which he had introduced into the science, he does not even mention the binominal nomenclature. In the *Systema Naturæ*, Ed. X, 1758, for the first time the binominal system is consistently applied to *all* classes of organisms (though it had been partially adopted by him as early as 1745), and hence many naturalists have regarded the Xth edition as forming the most natural starting point. The system being of slow and intermittent growth, even with its originator, an arbitrary starting point is necessary. In the XIIth edition (1766–68), numerous changes and reforms were instituted, and a number of his earlier specific names were arbitrarily changed. In fact, Linnæus never seems to have regarded specific names as subject to his rules.

It must be observed that an apparent rather than real distinction has been observed, especially by botanists, between the citation of the *authority* for names of genera and that relating to specific names. In the early part of the eighteenth century a few botanists, among whom Tournefort (Rei Herbar. 1749) may be particularly mentioned, had progressed so far as to recognize and name under the title of genera, groups corresponding in most essentials to the modern idea of genera. Linnæus himself adopted a number of these, and used the names of Tournefort and others as authorities after the generic name as adopted by himself. In this the great Swede has been almost unanimously followed by botanists, though the names take date only from the time of their adoption by Linnæus.

A very few authors, Bentham being the most prominent, have refused to cite any one except Linnæus as authority for such genera.

Whether the course of the majority be considered judicious or not, it is now the accepted usage in Botany. As regards names in general, botanists appear to agree in adopting the date of the Linnæan *Species Plantarum* (1753), as the epoch from which their nomenclature must begin. This work contains the first instance of the consistent use of the *nomen triviale*, subsequent to the proposition of the rules in the *Philosophia Botanica*, to which modern nomenclature is due.

Binomial designations cannot, of course, be reasonably claimed to antedate the period when binomial nomenclature, in a scientific sense, was invented, and in spite of the solitary instance of 1745, no good reason appears for extending the range of scientific nomenclature to an earlier date than 1751.

In 1842, the committee of the British Association reported " it is clear that as far as species are concerned we ought not to attempt to carry back the principle of priority beyond the date of the twelfth edition of the *Systema Naturæ*. Previous to that period, naturalists were wont to indicate species not by a name comprised in one word, but by a definition which occupied a sentence, the extreme verbosity of which method was productive of great inconvenience. It is true that one word sometimes sufficed for the definition of a species. but these rare cases were only binomial by accident, and not by principle, and ought not, therefore, in any instance, to supersede the binomial designations imposed by Linnæus." This related solely to zoology.

It is said that in the original draft of the report, the number of the edition of the *Systema Naturæ* was left blank, and afterwards filled up by the insertion of "twelfth." This insertion renders the paragraph, otherwise judicious and accurate, glaringly incorrect. What motive resulted in the selection of the twelfth edition as opposed to the tenth, or of any special edition after the adoption of the binomial form by Linnæus, has never been set forth in a satisfactory manner. If any special edition were chosen, the tenth has *prima facie* claims for first consideration. It is as clearly binomial as any, and it is as consistently so. The inference from the paragraph as it stands, that the *Mus. Tessinianum*, the tenth and eleventh editions of the *Systema*, the writings of Pallas, etc., were only accidentally binomial, is too preposterous not to create a reaction against the committee's recommendation in the mind of any one familiar with these works. This has been the result, in fact, and, to a considerable extent, in the works of the naturalists of northern Europe, the tenth edition has been taken as the starting point.

The rule following the paragraph quoted from the committee's report, however, contains no reference to any special edition or work, and reads, simply: — " the binomial nomenclature having originated with Linnæus, the law of priority in respect of that nomenclature, is not to extend to the writings of antecedent authors." (§ 2, p. 10, l. c.)

There is nothing to object to in this, and it is in all respects reasonable and fair. In adopting the tenth edition as a starting point, therefore, naturalists have not infringed on any of the original B. A. rules, but, at most, disregarded a subsidiary recommendation founded on an inaccuracy.

In 1863, however, the Association determined to overhaul the rules, and the amended ones adopted in 1865 contain the following remarks and modification of the second rule:

"The Committee are of opinion, after much deliberation, that the twelfth edition of the *Systema Naturæ*, is that to which the limit of time should apply, viz., 1766. But, as the works of Artedi and Scopoli. have already been extensively used by Icthyologists and Entomologists, it is recommended that the names contained in or used from these authors should not be affected by this provision. This is particularly requisite as regards the generic names of Artedi, afterwards used by Linnæus himself."

The original rule of 1842 is then modified (by the additions indicated in italics) to read as follows:

" The binomial nomenclature having originated with Linnæus, the law of priority, in respect of that nomenclature, is not to extend to the writings of antecedent authors, *and, therefore* (sic), *specific names published before 1766 cannot be used to the prejudice of names published since that date.*"

It would appear that the Committee were " plus saint que le Pape," since they would reject names that Linnæus himself was ready to and did adopt. In this connection, Prof. Verrill (Am. Journ. Sci. and Arts, July, 1869) has made some judicious remarks calling attention to the works of Pallas, and Thorell has done the same for Clerck on the subject of spiders.

An apologetic paragraph, following the remarks above quoted from the B. A. Committee report for 1865, inferentially admits the error of 1842, but goes on and reaffirms it on the ground that confusion would otherwise result.

It is very doubtful if much confusion would be caused by leaving the question open, since half the naturalists of Europe and America have already adopted the tenth edition of their own motion, and the other half, or a large portion of them may not unreasonably be believed to be only held back from joining the others by a desire to conform to the rules, even where injudiciously framed.

In a large part of zoology, the change would make no difference whatever, since the scientific study of such branches has begun since 1766.

The principle embodied in the Rule now submitted by your Reporter, is inferentially admitted to be valid by the B. A. committee in their remarks on Artedi and Scopoli.

Since 1865, Thorell, in his monograph of European spiders, has boldly adopted, so far as specific names[1] are concerned, a plan similar to that here recommended, and takes the binomial work of Clerck (1757) on Swedish spiders as his "epoch maker."

Geo. R. Gray, in his " Genera of Birds," adopts the first edition of the *Systema* (1735) much in the botanical sense as the epoch-maker for ornithological genera. For specific names he does not go behind the tenth edition (1758).

Alex. Agassiz, also, while declining to be bound by any arbitrary rules, has gone to the foundation of things in Echinological literature, and from the typical specimens of their describers, has brought into scientific nomenclature, the ancient and earliest names of Klein and others, whose work in this branch of zoology he pronounces far in advance of their time.

However inadvisable such changes would be in any department where the nomenclature may be tolerably well fixed; in those in which students are overwhelmed by

---

[1] The genera of spiders are not affected by the question.

the mass of undigested synonymy, as in Conchology and Entomology, it cannot be doubted that such work thoroughly and accurately done would be a vast relief.

For this reason I cannot but be impressed with the value which the compromise here suggested might have for various branches of zoology, should specialists look upon it favorably. It has already been suggested, with favorable comments by Dr. Leconte, that such a course should be pursued, and the suggestion has been carried still further in the rules presented for consideration to American entomologists, by Messrs. Leconte, Riley and Saunders.

Those departments having a tolerably well settled nomenclature, whether trembling in anticipation of sweeping innovations or not, would thus have it in their power to settle the matter once for all, and in the interest of the convenience of all concerned. Once settled, the subsidiary rectifications would be a mere matter of time.

It may be enquired, why, after criticising adversely the selection of the twelfth edition as a starting point, it is still recommended for retention in the event of specialists failing to agree on any other basis?

The reasons are,— 1st, because it has twice been recommended by the British Association committee, thus acquiring a certain status, and change is always undesirable unless it goes to the root of a matter and in a way to receive general assent; and 2nd, because in one of the branches most concerned in the settlement of the whole question, it appears for the stability of its nomenclature, to be a vital point that the usage founded on the B. A. rules should be maintained if possible; at least until the students of that branch (Entomology) shall decide on an epoch for themselves.

If it be insisted, in accordance with the old method that but one epoch shall be considered for all zoological nomenclature, it appears to the Reporter that the date of the tenth edition of the *Systema Naturæ* should be taken without regard to changes necessitated thereby.

More than half the replies to the circular are in favor of this starting point or one even earlier, the botanists of course insisting, in behalf of their department, on 1753·

§ LIX. In botany the epoch for scientific names is 1753, the date of the Species Plantarum of Linnæus.

The epochs chosen by Thorell for Spiders, and Agassiz for Echini, have already been alluded to.

Of Synonyms and Works subject to Citation.

§ LX. It is advisable in preparing tables of synonymy of a monographic character, to divide the citations into several groups according to their nature, the several citations in each group following each other in the order of their dates (which should always be specified) beginning with the earliest.

All synonymy is of a historical nature but may be divided into two groups of which one is strictly historical and the other biological. The historical series includes the citation of all authors who may have mentioned the organism and primarily exhibits the fluctuations of its nomenclature. It is divisible into two sections :—

1. *Pre-Linnæan* or *Bibliographical*, including citations of names in works antecedent to the epoch adopted for the starting point of the nomenclature of the class to which the organism belongs.

2. *Binomial*, containing citations of names from works in which the scientific nomenclature has been adopted.

The biological series contains citations of works in which additions have been made to our knowledge of the organization, development, distribution or genetic relations of the organism concerned.

While the *names* used in the biological series may properly be cited in the historical series, authors of monographs may greatly facilitate the investigations of students by appending to their historical synonymy a biological list, especially, if, in the latter, they add to the citations of the papers or publications themselves, a word or two in parentheses indicating the character of the additions to knowledge to be found in the respective works,— as (Embryology), (Geogr. Distr.), etc.
The advantages of this course are sufficiently obvious.

§ LXI. Synonymy, properly speaking, has reference only to names coming under the second section of the historical series. Its object is solely the sifting of the nomenclature of an organism or group of organisms, that the name or combination of names which are entitled to be permanently connected with it, and by which it shall be denominated in scientific literature, may be definitely determined.

The object of citations in the biological series is of a different character. They are intended to serve as a guide to the student in researches of a biological nature as an index to the progress of investigation, and to the views on the subject held by different authorities. As such, the series should be disembarrassed from extraneous, purely synonymical citations. This distinction, especially noted by A. Agassiz, has hitherto been much neglected.

§ LXII. The following kinds of works are entitled to citation in bibliography, but not in synonymy.

1. Works antecedent to the nomenclature-epoch adopted for the class of organisms concerned.

2. Works subsequent to that epoch in which the binomial nomenclature is not consistently adopted.

3. Works not published.

Works of the first and second kind may often be advantageously consulted for biological information and cited in the biological series; as, for instance, Poli, who adopted a singular quadrinomial form of nomenclature, but to whom are due many important anatomical researches. No injustice is done by citing an author for the real benefit he has conferred on science, while declining to burden the latter with an incompatible nomenclature.

With regard to works of the second kind much diversity of opinion has been expressed. Some naturalists would accept in synonymy, accidental binomial phrases, as entitled to priority as names, provided they occur in works published subsequently to the nomenclature-epoch. But in this as in all cases when tested by fundamental principles it is easy to arrive at a conclusion.

When one word at the commencement of a descriptive sentence or phrase is in italics or separated by a comma from what follows (as was the usual practice amongst ancient writers) it is a matter subject to a diversity of opinions as to whether this, together with the generic name if any were employed, forms a tenable binomial name or not. Some authors would adopt the genus and consider that no tenable specific name had been employed. Others would adopt both. Others again would reject both. But the question as to whether the author in the work referred to, adopted and consistently used the binomial system of nomenclature is one upon which no difference of opinion can exist. It is a question capable of a categorical answer at once. Hence it would seem preferable to stand, once for all, on the solid ground of certainty and avoid a course which will inevitably introduce a large amount of uncertainty. This course is that recommended by the B. A. Rules and already adopted by a majority of naturalists.

It may seem superfluous to object to works of the third category. But beside several MSS. preserved in Museum libraries and sometimes quoted, though never printed, there are a few works which have been printed but never published. This is the case with a Museum Catalogue prepared by Link about 1806.

It was printed and contained a host of new names. But whether the author was ashamed of his work, or the authorities of the University declined to be sponsors for the innovations, the work was never offered for sale, distributed, or advertised by the author.

Only one copy is definitely known to have escaped from the University cellars, and it has been stated that the remainder, or most of them, were destroyed by fire. Yet in 1851, the solitary copy having been discovered, one or two authors called attention to it and demanded that these names should take precedence of those of Lamarck and others, which had been in use for nearly half a century. A few writers have adopted this suggestion, and in one branch of science at least, deplorable confusion has resulted.

The auctioneer's catalogue of Bolten's collection printed in 1798, but fortunately containing no diagnoses, and of which only one or two copies are known, falls nearly in the same category. A reprint was made in 1819 but is also one of the rarest of books.

§ LXIII. To avoid increasing the difficulties encountered in dealing with the already enormous mass of scientific names, authors are earnestly recommended to take the following precautions in publication.

1. To publish matter containing descriptions of new groups or species in the regularly appearing proceedings of some well established scientific society, or in some scientific serial of acknowledged standing and permanence.

2. If a separate publication or independent work be issued by any author, copies should at once be sent to the principal learned societies, scientific libraries and especially to those persons or associations known to be engaged in the publication of bibliographical records or annual reviews of scientific progress.[2]

The work should also be placed at the disposition of the scientific world by an advertisement of copies placed in the hands of some firm, society or individual, for sale or distribution.

3. To avoid most carefully the publication of new names or changes of nomenclature in newspapers; serials not of a scientific nature or of limited circulation; in the occasional pamphlets issued by weak, torpid or obscure associations which are distributed only to members or not at all; and in brief lists, catalogues or pamphlets independently issued, insufficiently distributed or not to be found on sale.

The question of the restriction of the nature of the channels through which additions to, or changes in nomenclature may be made, so as to exclude from consideration in synonymy such publications as do not conform to the proposed restrictions, has frequently been mooted. It was touched on in your Reporter's circular (XXV, XXVI) but the replies were almost unanimously to the effect that however desirable it may be, the plan is impracticable; a judgment in which the Reporter, with some reluctance feels obliged to concur.

It is clearly, however, the duty of every publishing author to concur as far as pos-

[2] This can at present easily be done with hardly any expense, in this country, through the Smithsonian Institution and its agents.

sible in the suppression of methods leading to confusion. The above recommendations are intended to lead toward this result. In this era of numerous serials of high
scientific standing as well as learned societies established on a permanent basis, there
does not seem to be any good reason for failing to comply with such regulations as
may be best fitted to advance the convenience and best interests of all.

§ LXIV. The majority of naturalists interested in the study of any
special class of organisms may properly unite in expressing an opinion in
regard to any particular work treating of those organisms, in regard to
the sufficient publication of which, at a certain date, doubts may exist;
as to whether said work be entitled to be quoted in synonymy as well as
in historical and biological bibliography.

It is, therefore, recommended, that in such cases should a decisive expression of opinion seem necessary or desirable for the benefit of science;—1st, that due notice of the proposed action be previously given to
those interested; 2nd, that when the action has been had that the results
should be as widely published as possible; 3rd, that the decision of the
majority, once made, be concurred in cheerfully by all the naturalists interested, for the common good.

Almost every department of zoology, within the last twenty-five years, has suffered
from the discovery of some utterly forgotten work, in which names had been applied
to objects subsequently made known to the scientific world under other appellations,
which latter have been generally adopted. Such obsolete works are generally of no
intrinsic scientific value, and are only recalled to public notice by the action of the *lex
prioritatis* on the names they may contain. Even when at the time of their production
they formed an advance on the knowledge of that day, it is inevitably the case at present that this advance is immensely behind the present state of the science, so that
while the restoration of the obsolete names is in accordance with the principles which
rightfully govern nomenclature, the effect upon biological study (of which nomenclature is only one of the conveniences), is positively harmful. Names sanctioned by
years of usage, common to the whole literature, having a traditional as well as a present value, and standing for certain verified conclusions,—are replaced by unfamiliar
terms, which are perhaps conceived in error, foreign to modern scientific literature,
and too often requiring the transposition of other, and familiar names from groups in
connection with which they are universally known perhaps to groups equally well
known under some other name,—thereby producing the most lamentable confusion.

It is evident that the utmost care must be taken in dealing with the subject, since
any general rule intended to arbitrarily discriminate, would be only too liable to react
injuriously on other portions of nomenclature, in a way to make the justice of the rule
questionable, bring its authors and supporters into disrepute, and, by failing to induce
uniformity, to aggravate the difficulty it was intended to overcome.

Many naturalists have given much thought to the subject and various propositions
have been suggested as a remedy. One of these which seems most reasonable, and
which a number of naturalists agree in supporting, is to the effect that a name which
has not been in use for (say) twenty-five years, shall be excluded from use thereafter
in that special connection, as by a statute of limitations. (Riley, Lec.) Others, as Mr.
Lewis, desire that only such names as are actually "in use" at a certain period shall
be retained.

The objection to the first method is that it is arbitrary, that it will not necessarily
produce permanency (for some other work just within the limit of time may turn up
and reverse the decisions founded on the supposition that a certain name had not been
"in use" within the period), and especially, as also in the second case, that the term
"in use" is not susceptible of an exact definition. The proposition of a name by an
author may or may not cause it to be "in use." It may be overlooked or rejected by

others, or without for some time appearing in any other publication. it may be in very general use in letters, museum labels, etc. Everybody who is guided in his work by the particular views of the author referred to, in one sense "uses" the name though he may never publish it on his own account. No one can ever be sure that a contemporary name, even if glaringly erroneous, may not be (by the ignorance or error of some provincial naturalist) accepted in various obscure publications.

Hence it would appear that these propositions are impracticable, and this is the opinion of a very large majority of the American naturalists who replied to the circular in which the question was submitted to them (XXI, XXII).

The only other resource would then appear to be either (1), the publication of a list of names in any special group with their types, which all the specialists of that group would agree to accept, or (2), some such course as the one suggested by the foregoing § LXIV.

The first is clearly impracticable, since perfection in nomenclature, as in other things, is a relative term, and but a small proportion of the names could be settled with certainty, from the constant progress of our knowledge of the relations of organisms.

The second seems to be the only practicable method. The difficulties which all naturalists deplore, in the main arise, in each specialty, from some one or two obnoxious works. The course proposed by Alfred Wallace (Trans. Entom. Soc., London, 1871, LXVIII), recommended by Dr. LeConte (p. 210), and acted on in the " Rules to be submitted," etc., of Messrs. LeConte, Saunders, and Riley (¶ IV), in respect of this particular case, seem to your Reporter to indicate the proper method of cutting the Gordian knot and to be well worthy of consideration.

It should, however, be clearly borne in mind that this process is one that requires the greatest impartiality and delicacy in its application, and can only be justified by paramount necessity in any particular case. It is not in consonance with some of the general principles of nomenclature, can only be accepted when applied with self-evident propriety, and any license or abuse of the power it implies would react against the authors and make the last state of the science worse than the first.

Of Names to be Rejected, Changed, or Modified.

§ LXV. A name cannot be changed under the pretext that it is badly chosen, that it is not agreeable, that another is better or more widely known, that it is not of a sufficiently pure Latin derivation, or for any other contestable or valueless motive. (DC.)

§ LXVI. A name should be rejected under the following circumstances.

1. When this name has previously been applied in a tenable manner to another valid group of organisms in the same kingdom. (DC., Lec., B. A.)

2. When it is already applied to another species in the same genus, or to another subdivision of a species in the same species. (DC., B. A.)

3. When it expresses an attribute or character positively false in the majority or the whole of the group in question, as in cases (among others) when a name has been founded on a monstrous, abnormal, immature, artificial or mutilated specimen ; or when a geographical specific name is that of a country entirely removed from the faunal or floral grand division to which the species belongs. (DC., B. A., Th., V., Bourg.)

4. When it is formed of two words belonging to different languages, as *eu* put before a Latin name, *sub* before a Greek name, *oides*, *opsis*, suffixed to a Latin name, etc. (DC., Th., etc.)

5. When it is contrary to the provisions of §§ L,-LIX.

6. When it was published in a work not entitled *a priori* to be cited in synonymy (§ LXII), or which has been definitely excluded from synonymy by such action as is suggested in § LXIV.

7. When it outrages decency or religion. (Lec. 202, ¶ 5.)

It has happened that persons of scientific acquirements but of unbalanced mind or depraved taste have applied grossly indecent or blasphemous names to sundry organisms. It would seem as if a more thorough punishment could hardly be devised than the permanent attachment of their names as authority to the appellations referred to. But, since these names would remain a stigma on science as well as on their originators, and would outrage the feelings of a majority of naturalists, it is probably better that they should be suppressed. A few old names, originally in bad taste, have become so identified with the organisms they indicate, as to have ceased to be seriously offensive, and, being consecrated by usage, need not be disturbed.

8. When the name has never been defined, and a properly defined name has subsequently been applied to the same group, the earlier name should be rejected.

If the earlier name be known to the describer he will do well to define and adopt it when it will be accredited to him as authority, thus saving a synonym, but when this course has not been adopted it is too late to establish the undefined name.

9. When a name belonging to the Latin or to a modern language, and having an unquestionable and specific signification, is applied to a group which it cannot by its etymology properly indicate it must be rejected. (Th.)

Thus *Tarantula* Fabricius (1793), not being the historic tarentula is properly rejected for the later *Phrynus* Oliv., and the former name is applied to the true tarentula of the ancients. (Th.) But if another name had been the first applied to the true tarentula in scientific nomenclature it could not have been suppressed to revive the classical name. There are very few instances where the species can be identified with sufficient certainty to justify the rejection of a name on the above ground as held by Thorell.

10. When a name is identical, when properly spelled according to a derivation given by its author, with a prior valid name in the same kingdom, it must be rejected. (Th.)

This is sufficiently obvious.

§ LXVII. When a name has been used in one kingdom subsequent to 1842, it should thenceforth be ineligible for use in the same kingdom except within the same order (ordo) in which it was originally applied.

It would be better for Science if all names which have once been used should hereafter be ineligible in the same kingdom. (Th., Bd.)

But previously to the promulgation of the B. A. rules little intercourse existed between the naturalists of different countries compared with what now obtains. So it has happened that the same name has been used several times for different groups, and in some of these cases being certainly invalid, has been applied to the first valid group for which it had been proposed. In this manner it may have come into general use and to reject such names in some cases would cause confusion.

It may, therefore, be considered as too late to propose a radical change, though by Baird, in ornithology, and others, this principle has been carried to its conclusions. But for more modern names the adoption of some such method seems urgently called for.

It might, at first sight, seem as if the reverse of the rule proposed would be desirable, i. e., that the name, if used again at all, should be valid only in *another* order than that in which it was first proposed. But, in the great multiplication of genera brought about in recent years, many are necessarily tentative, and depend for their adoption on the judgment of the best authorities for the particular order. Some may hold a genus valid, others reject it. It is not reasonable that students of one group should be kept waiting until a definite decision is arrived at (even if that were possible) in another group with which they are not familiar, and until the students of the latter group have decided whether to adopt or reject a name which has been used in both orders. (Scudd.)

Shall the student of Brachiopods wait until the Dipterologist shall decide the value of Robineau Desvoidy's genus *Megerlia*, before the former shall be able to adopt or definitely reject *Megerlia* of King? It has not been done in forty-five years, and *Megerlia*, King, is still in the limbo of names without a clear title. This is undesirable and wrong. Let the student of flies decide for himself whether he can retain Desvoidy's name or not, but let the mere fact that it has been used in a different class (classis) enable the student of brachiopods to reject it and place his own terminology on a sound foundation.

Upon the validity of the genus *Pelagia*, Peron and Lesueur depends the right to exist of eight or nine other generic names in different classes. By the method usually tolerated, these questions can never be settled. Let all, except the student of Acalephs, then, have the right to wipe out the name, *Pelagia*.

It would be vastly better if this principle could be applied to all names; but it will be at least a partial and important relief to be permitted to apply it to those of modern date.

§LXVIII. When an inelegant combination has been the result of using for the name of a new genus, a name or a modification of a specific name borne by a species which is to be contained within the new genus, the fact is not a sufficient reason for rejecting the generic or changing the specific name. (L., Ag., Th., etc.)

This necessarily follows from § LXV, but it seems desirable definitely to state it.

The practice is objectionable on account of its producing tautological inelegance, and because it has resulted in the formation of a number of generic names of adjective form.

On the other hand in connection with certain of the Linnæan and other ancient and universally known species, it had several beneficial effects. It recalled the typical form for which the genus was constituted, and in many cases it might rightly be regarded rather as a change of rank than the creation of a new name. The ancient species (ex. *Voluta oliva*, Lin., genus *Oliva* Brug.) often covered an assemblage of forms equivalent to a modern genus. A vast number of the old names were thus constituted by Linnæus, Lamarck, Cuvier, Agassiz, and the fathers of science.

Their practice was to replace the old specific name by a new one. This practice was reaffirmed by the B. A. committee in 1842, as follows:—"§ 13. A new specific name must be given to a species, when its old name has been adopted for a genus which includes that species." The usage was very properly condemned by the Committee, as its benefits do not extend to little known modern specific names, while the objections to it are as forcible as ever.

But in 1865 the Committee decided that the usage of a century must be reversed, and the following modified rule was the result.

"§ 13. A specific name must not be altered to use that name for the genus; where this has been already done the old specific name must be restored, and a new generic name given to prevent an inharmonious repetition."

This innovation, the sweeping character of which the Committee cannot have realized, if carried into effect would uproot hundreds of the generic names best known to science, and so familiar that the fact that they were originally specific names has been almost totally forgotten. Its spirit is opposed to the fundamental principles of nomenclature, and the end to be gained is of the most trivial character. The gain in elegance is not apparent in the substitution of *Crassivenus (!)* for *Mercenaria*, or *Tottentana (!!)* for *Gemma*, as generic names. But a few naturalists have followed the new method, and the replies to a question (XI) on the subject included in your Reporter's circular have been practically unanimous in favor of the rule as here stated.

In cases where the specific name was changed by the author of the genus and has come into general use, no good end will be gained by attempting to revert to the now forgotten original specific name. The retention of the latter is recommended merely for such cases as have given rise to controversy, or where the substituted specific name has not come into use. It is a singular fact that several of the more glaring instances cited as examples of inelegance by those in favor of the innovation show, when investigated, that the circumstances have not been understood. Thus, the original generic name of *Gemma*[3] *gemma* Deshayes was not *Gemma*, and neither the generic nor the specific name of *Gari gari* Schumacher are tenable; since the first is improperly formed and should be *Garia*, while the latter was applied by Rumph, a non-binomial author, and is a vernacular and not a Latin or Latinized term. But in this, as in other cases, the usage of a hundred years and the fixity of nomenclature will outweigh any consideration of mere elegance.

Of Changes of Names.

§ LXIX. An author has no rights in regard to the change or rejection of names of his own proposition, except those which are common to all naturalists, and authorized by the rules of nomenclature. (DC., Th.)

When a name has not come into use, and its originator proposes a change not contrary to the spirit of nomenclature, it may, out of courtesy, be adopted by others and by usage become justified, even when not directly authorized by the rules. But when such a change is in direct opposition to the rules it cannot be sustained. An author has the same rights, no more and no less than other naturalists, over names he himself has proposed. In effect publication is a fact which cannot be annulled. (DC.)

§ LXX. The name of a cohort, subcohort, family or subfamily, tribe or subtribe, should be changed when the genus, from whose name it is derived, is known to no longer form part of the group in question. (DC.)

§ LXXI. When a section or subsection of a genus, preserving its rank, is transferred to another genus in which there is already a subordinate group bearing the same title, the first mentioned name should be changed. (DC)

§ LXXII. When a species is transferred from one genus to another, in which latter there is already a species of the same name, the newest of the two identical specific names should be replaced by its first tenable synonym, or by a wholly new specific name if there be no synonyms.

When a species has received a new name by reason of the existence of a prior valid species of the same name in the same genus, and this second specific name has become fixed by usage, it is not necessary in subsequent transfers to recur to the first rejected specific name, unless some beneficial end is to be attained.

---

[3] Possessing Deshayes original MSS., I am able to say that his name was *Gemmula* and not *Gemma*, which latter was a typographical error. He did not see the proofs.

To illustrate this by a suppositious example let us assume that a *Buccinum striatum* was described by Linnæus; that Lamarck subsequently described a *Buccinum striatum* whose name was changed by Wood, on account of the existence of the Linnæan species, to *B. rugosum*. *B. rugosum* is found to belong to the genus *Nassa*. Yet it is not imperatively necessary to change the name *Nassa rugosa* (Wood), to *N. striata* (Lam.), after it had come into use. But if there had been already a *Nassa rugosa*, even if of later date than Wood's description, it would then be better to reinstate the name *striata* Lam., and thus confine all the changes to one species instead of embroiling two.

§ LXXIII. When a subspecies, variety, or other subdivision of a species, is referred to another species, the name of the former should be changed, if the latter already contains a modification of the same rank or kind bearing the same name. (DC.)

§ LXXIV. When a group is transferred from that which contains it to another, and the first preserves its original rank, its name should be changed if it becomes senseless, or an evident cause of error and confusion in the new situation in which it is placed. (DC.)

§ LXXV. In the preceding cases the name to be changed or rejected, is to be replaced by the oldest valid name of the group concerned; or, if there be none, a new name should be created. (DC.)

§ LXXVI. Names of groups having higher than generic rank may have their terminations modified to render them conformable to usage or the rules. (DC.)

§ LXXVII. When a name (1) stated by the author to be derived from certain Greek or Latin words exhibits a glaringly faulty construction; or (2) when its orthography is evidently erroneous; or (3) when a name taken from the name of a person has not been written according to the real orthography of his name; or (4) when a wrong gender has been attributed to a name by its termination, either by its not agreeing in gender with the genus to which it is referred, or from some internal inconsistency in the name itself:— Naturalists are authorized to correct the faulty name or the faulty termination, provided the name be not very ancient and universally received under the erroneous form. This authority, however, should be used with great reserve, particularly if the change will effect the first syllable, and above all the first letter of the erroneous name. (DC., Th., B. A.)

1. Some naturalists would leave unchanged all errors of construction, no matter how glaring and correct only orthographical errors, but by far the greater number including the B. A. committee regard serious errors of both kinds as subject to reformation. Mongrels or compound mutilates are generally rejected (Th.) when they cannot be reformed. *Valenciennensis* Rousseau, intended for a generic name, has properly been changed to *Valenciennesia* by Fischer.

2. In Latinizing Greek words there are certain rules of orthography which should never be departed from. (B. A.) Hence such names as *Aipucnemia, Zenophasia, Agkistrodon,* must, according to the rules of etyomlogy (see § LXXXIV), be written *Æpycnemia, Xeniphasia, Ancistrodon,* and have accordingly been changed.

3. For instance, *Mangilia* Risso, derived from the name of Mangili, an Italian naturalist, has been most generally written erroneously *Mangelia.*

But to change certain universally known names on account of erroneous orthography would be very inconvenient. For instance, in the Botanical Congress at London, in 1866, it was proposed to modify the name *Cinchona*, on the ground that it was named after the Countess Chinchon, but the majority of the botanists present, advised the continuance of the established usage.

Gundelia is very far from Gundelscheimer, but since the ancient botanists permitted this license and it has been consecrated by an hundred years of habitual use, why change it? The purists alone remember Gundelscheimer, and the name *Gundelia* has been accepted as an arbitrary name. In these questions it is necessary to bear in mind that the fixity of nomenclature is the interest of greatest importance, and that a naturalist has the right (under certain limitations) to construct a generic name in any manner whatever; for example, in a way which may resemble the name of a person. (DC.) Vernacular names will be hereafter considered.

4. Thus, a *Scalaria* named after Miss Turton is to be written *S. Turtonæ* and not *S. Turtoni*. Names of nymphs and goddesses are necessarily feminine, and when used as generic names their terminations and those of their adjective specific names must also be feminine. *Viviparus* Montfort, is inconsistent with itself and must be written *Vivipara*.

To change the first syllable, above all the first letter of a name is inconvenient on account of the arrangement of indices, tables, catalogues and dictionaries in alphabetical order. It is very incommodious for instance, that several generic names commencing with E have been altered to He, on account of the rough Greek accent. These names have to be sought for in two places in all the tables. The Greek accents varied according to the dialects; it is hard to see why naturalists should be more rigorous than the Greeks. (DC.)

Names of uncertain etymology should not be changed on etymological grounds. (Th.)

§ LXXVIII. Although names of persons are Latinized and not adopted in a Greek form, no one is authorized to reject or change the name of a group compounded from that of a person with the prefix *eu*, *oides* or *opsis*. It is sufficient that they are not of Latin origin, for it is, above all, necessary to avoid changing names, but a naturalist of taste will avoid the origination of such.

Some naturalists have amused themselves by applying to species names which are Latin puns on the names of persons. *Trochus faba* McGillivray, ostensibly named after Mr. Bean; *Fusus domænovæ* Val., after M. Maison-neuve, are examples. These names, however absurd, should not be changed, for, with the motives of the author of a name, one has nothing to do, provided the result, considered in itself, is not seriously objectionable.

§ LXXIX. When the name of a person is used without the necessary Latinization of the termination, unless it already possesses a form similar to that provided for under § XXII, it must be changed to conform to the rules.

Example: *Clausilia Mortillet*, Dumont, is properly *C. Mortilleti*. Some naturalists have even gone further, and proposed specific names composed of two words, the Christian and surnames of the person to whom the species was dedicated, and without Latinization! Example: *Donacia Scipion Gras*, Mort., after M. Scipion Gras as distinguished from his brother, Albin Gras, for whom another species of *Donacia* was similarly designated. Such names are deplorable and cannot be accepted. (Bourg.)

§ LXXX. When a name is derived from a vernacular word it must be accepted in the form given to it by its original describer, even in cases where the orthography of the name has been insufficiently understood by the author, and has given rise to merited criticism. (DC.)

Vernacular names, above all, when taken from barbarous languages, are frequently uncertain, and the manner of writing them doubtful. When once the name has been adopted by science, it would be too easy to change it, if one professed a rigorous exactitude. (Cf. for exceptions, note 3, § LXXVII) *Coffea*, for example, would become *Cavea, Covea, Cauffea*, etc., according to each person's idea of the orthography of. the Arabic name. The same principle holds good in relation to species allied or distinct, but having in different localities the same vernacular name derived from some character common to them all. A naturalist fixes the name on some one of the species, no matter which; otherwise it would be continually contested or changed. (DC.)

§ LXXXI. Vernacular names (1) derived from classical names by a light modification, may, if unmistakable, be changed by subsequent authors, into their proper Latin forms, but are not entitled to priority if not so adapted before another name properly formed has been applied to the organism in question. (Rep.)

Vernacular names (2) not so derived have no standing in science, and if a classical term having the same meaning is subsequently applied to the same organism or group, its priority dates only from the time when this was done by some author, who is entitled to be cited as authority for the name. (B. A. Bourg.)

1. *Patelle viride* may be supposed to stand for *Patella viridis*, and, properly modified by courtesy, is liable to be accepted, if no Latinized name precedes the change in time of application.

2. No one is obliged to suppose that, for instance, *Vis aguillée* represents *Terebra aculeata*, until some author applies the latter name to the mollusk in question. (Bourg.)

§ LXXXII. Vernacular names, especially when applied to genera and species, if not proper nouns or having already an accidental Latin form, should be changed to conform to the rules of Latin orthography. (B. A.)

A pernicious practice, of very old date, exists, of applying to species, names not only of barbarous origin, but without Latinization, and totally destitute of euphony. These are chiefly the local appellation of some savage tribe for the organism designated.

Thus we have *Hyperoödon butzkopf*, Gray, *Balœna tschielagliuk* and *B. agamachtschik* Pallas, etc. *Butzkopf* is an obscene name applied by the Dutch boers of South Africa to any species of whale, though supposed by Gray to be a local name for the species he indicated, and the two species of Pallas, with others of the same kind, are described from Kamchatka, where the names he used are unknown.

The practice has probably gone too far, so far as proper nouns are concerned, to admit of the total rejection of such names when they are known to apply to the particular species referred to (§ LXXX). But when these barbarisms are adjective in their nature, they are on the principles herein before cited, subject to reformation as pointed out by the B. A. Committee. No naturalist of good taste would inflict such names, unmodified, upon science, and it would be far better if they were totally avoided.

Of the Names of Organisms in Modern Languages.

§ LXXXIII. Naturalists employ in modern languages scientific Latin names or those immediately derived from them in preference to names of other origin or kinds. They avoid using these latter, except when they are very widely known and clearly intelligible. All friends of Science should oppose the coinage of new vernacular names not existing in any language, at least when they are not derived from the scientific Latin name by means of some slight modification. (DC.)

It is desirable to continue the use of Latin for scientific descriptions, and still more for names. These, like proper names of persons, should be available in all languages. The public adopts them rapidly as a matter of habit, even when they are grotesque. No one objects to such names as *Fuchsia, Rhododendron,* etc., now common to all countries. For much stronger reasons, it is desirable to proscribe the fabrication of so-called "common names." The public to whom they are addressed finds no advantage in them, for they are new to it, not to be found in any dictionary, and the learner must also know the Latin name of which the new term is a synonym. To comprehend how confusing a multiplicity of "common" names would be, it is only necessary to suggest, for instance, the condition of the Post Office department, if each village had an absolutely different name in each language. (DC.)

The manufacturers of new "common" names occasionally make some amusing slips, as in the case of a recent attempt in England, when for several groups of animals which already possess well known and characteristic names in the vernacular, new "common names" were proposed.

Of the Latinization of names.

§LXXXIV. Greek names are Latinized by substituting for the Greek letters their Latin equivalents according to the following table. (Herrm., Th., B. A., etc.)

α	= a ;	(βῆτα)	Beta.
β	= b ;	(βραχίων)	Brachium.
γ	= g ;	(γλῶσσα)	Glossa.
δ	= d ;	(διψάς)	Dipsas.
ε	= e ;	('υαλέος)	Hyalea, not Hyalaea.
ζ	= z ;	(ζίζυφον)	Zizyphus, Zizyphinus.
η	= e ;	(πειρήνη)	Pirena, not Pirina.
η final	= a ;	(πειρήνη)	Pirena, not Pirene.
ϑ, θ	= th ;	(τηϑύς)	Tethys; (θέτις) Thetis.
ι	= i ;	(βαλιός)	Balia, not Balea.
χ	= c ;	('ιπποχρήνη)	Hippocrena, not Hippochrenes.
λ	= l ;	(φυλλίς)	Phyllis.
μ	= m ;	(μέλας)	Melas.
ν	= n ;	(πειρήνη)	Pirena.
ξ	= x ;	(ξένος)	Xenus, Xenophora.
o, ω	= o ;	(φορός)	Phorus; (πῶμα) Poma.
π	= p ;	(ποταμός)	Potamus.
ρ	= r ;	(πτερόν)	Pterum.

ρρ	= rrh;	(φυλλίρ-'ροή)	*Phyllirrhoa*, not *Phyllirhoë*.
σ, ς	= s;	(γλωσσός)	*Glossus*.
τ	= t;	(πτερόν)	*Pterum*.
υ	= y;	('υβός)	*Hybolithus*, not *Hibolites*.
φ	= ph;	(φορός)	*Phorus*.
χ	= ch;	(κοχλίας)	*Cochlias*.
ψ	= ps;	(ψάμμος)	*Psammus*.
αι	= ae;	(λιμναῖος)	*Limnaea*, not *Limnea*.
αυ	= au;	(γλαυκός)	*Glaucus*.
ει	= e;	(τείνω)	*Exotenobranchia*.
εῖ	= i;	(χεῖλος)	*Chilostoma*, not *Cheilostoma*.
ευ	= eu;	(εῦρος)	*Eurus*.
ῳ, οι	= oe;	(δίς, οἰκέω)	*Dioeca*, not *Dioica*.
ον final	= um;	('εφίππιον)	*Ephippium*, not *Ephippion*.
ος final	= us;	('ομφαλός)	*Euomphalus*, not *Euomphalos*.
ου	= u;	(λουτήριον)	*Luterium*, not *Lotorium*.
γγ	= ng;	('αγγαρεία)	*Angaria*.
γχ	= nch;	(ἄγχω στόμα)	*Anchistoma*, not *Angistoma*.
γκ	= nc;	(ἄγκιστρον)	*Ancistrodon*, not *Agkistrodon*.
'ρ	= rh;	('Ρέα)	*Rhea*.
'	= h;	('ερμαία)	*Hermaea*, not *Ermaea*.

It may be remarked in concluding that, while the advances of philology have introduced reforms into the pronunciation and phonetic spelling of Latin words during recent times, these reforms cannot be allowed to affect words which, having obtained an entrance into nomenclature, have assumed, so to speak, an arbitrary character and signification. Otherwise new and annoying elements of confusion would be added, where there are already too many. The Latin of scientific nomenclature is, and must remain essentially the Latin of the eighteenth century.

www.ingramcontent.com/pod-product-compliance
Lightning Source LLC
Chambersburg PA
CBHW022022190326
41519CB00010B/1569